Beyond the Quartic Equation

R. Bruce King

Beyond the Quartic Equation

Birkhäuser
Boston•Basel•Berlin

R. Bruce King
Department of Chemistry
University of Georgia
Athens, GA 30602

Library of Congress Cataloging-In-Publication Data

King, R. Bruce
  Beyond the quartic equation / R. Bruce King.
    p.   cm.
  Includes bibliographical references.
  ISBN 0-8176-3776-1 (hardcover : acid-free). -- ISBN 3-7643-3776-1
  1. Quintic equations.   I. Title.
  QA215.K48   1996
  512.9'42--dc20   96-22841
                                                CIP

Printed on acid-free paper
© 1996 Birkhäuser Boston   *Birkhäuser*

Copyright is not claimed for works of U.S. Government employees.
All rights reserved. No part of this publication may be reproduced, stored in a retrieval system, or transmitted, in any form or by any means, electronic, mechanical, photocopying, recording, or otherwise, without prior permission of the copyright owner.

Permission to photocopy for internal or personal use of specific clients is granted by Birkhäuser Boston for libraries and other users registered with the Copyright Clearance Center (CCC), provided that the base fee of $6.00 per copy, plus $0.20 per page is paid directly to CCC, 222 Rosewood Drive, Danvers, MA 01923, U.S.A. Special requests should be addressed directly to Birkhäuser Boston, 675 Massachusetts Avenue, Cambridge, MA 02139, U.S.A.

ISBN 0-8176-3776-1
ISBN 3-7643-3776-1

Camera-ready text prepared by the Author in Microsoft Word.
Printed and bound by Quinn-Woodbine, Woodbine, NJ.
Printed in the U.S.A.

9 8 7 6 5 4 3 2 1

Contents

Preface vii

1. Introduction 1

2. Group Theory and Symmetry 6
   2.1 The Concept of Groups 6
   2.2 Symmetry Groups 9
   2.3 Regular Polyhedra 13
   2.4 Permutation Groups 19
   2.5 Polyhedral Polynomials 26
   2.6 Transvectants of Polyhedral Polynomials 30

3. The Symmetry of Equations: Galois Theory
   and Tschirnhausen Transformations 34
   3.1 Rings, Fields, and Polynomials 34
   3.2 Galois Theory: Solubility of Algebraic Equations by Radicals . 45
   3.3 Tschirnhausen Transformations 51

4. Elliptic Functions 56
   4.1 Elliptic Functions by the Generalization of Radicals 56
   4.2 Elliptic Functions as Doubly Periodic Functions 62
   4.3 Theta Functions 72
   4.4 Higher Order Theta Functions 79

5. Algebraic Equations Soluble by Radicals 82
   5.1 The Quadratic and Cubic Equations 82
   5.2 The Quartic Equation 87
   5.3 Special Quintic Equations Solvable by Radicals 89

6. The Kiepert Algorithm for Roots of the General Quintic Equation . 95
   6.1 Introduction 95
   6.2 The Tschirnhausen Transformation of the
       General Quintic to the Principal Quintic 99
   6.3 The Tschirnhausen Transformation
       of the Principal Quintic to the Brioschi Quintic 103
   6.4 Transformation of the Brioschi Quintic to the Jacobi Sextic . 108
   6.5 Solution of the Jacobi Sextic
       with Weierstrass Elliptic Functions 113
   6.6 Evaluation of the Weierstrass Elliptic Functions
       Using Genus 1 Theta Functions 118
   6.7 Evaluation of the Periods of the Elliptic Functions
       Corresponding to the Jacobi Sextic 123

6.8 Undoing the Tschirnhausen Transformations 126

7. The Methods of Hermite and Gordan for Solving the
   General Quintic Equation 128
   7.1 Hermite's Early Work on the Quintic Equation 128
   7.2 Gordan's Work on the Quintic Equation 132

8. Beyond the Quintic Equation 139
   8.1 The Sextic Equation 139
   8.2 The Septic Equation 143
   8.3 The General Algebraic Equation of any Degree 147

Preface

One of the landmarks in the history of mathematics is the proof of the nonexistence of algorithms based solely on radicals and elementary arithmetic operations (addition, subtraction, multiplication, and division) for solutions of general algebraic equations of degrees higher than four. This proof by the French mathematician Evariste Galois in the early nineteenth century used the then novel concept of the permutation symmetry of the roots of algebraic equations and led to the invention of group theory, an area of mathematics now nearly two centuries old that has had extensive applications in the physical sciences in recent decades.

The radical-based algorithms for solutions of general algebraic equations of degrees 2 (quadratic equations), 3 (cubic equations), and 4 (quartic equations) have been well-known for a number of centuries. The quadratic equation algorithm uses a single square root, the cubic equation algorithm uses a square root inside a cube root, and the quartic equation algorithm combines the cubic and quadratic equation algorithms with no new features. The details of the formulas for these equations of degree $d(d = 2, 3, 4)$ relate to the properties of the corresponding symmetric groups S_d which are isomorphic to the symmetries of the equilateral triangle for $d = 3$ and the regular tetrahedron for $d = 4$.

Related ideas can be used to generate an algorithm for solution of the general algebraic equation of degree 5 (the quintic equation). Such a quintic equation algorithm does not violate the classical theorem proved by Galois since it contains more complicated mathematical functions than the radicals that suffice for the algorithms for roots of the general quadratic, cubic, and quartic equations.

The underlying mathematics for an algorithm to solve the quintic equation was developed by nineteenth century European mathematicians shortly after Galois' discovery of the insolubility of the general quintic equation using only radicals. The initial work in this area was done by Hermite and then developed further by Gordan. This culminated in two key publications, an 1878 article by Kiepert[1] describing a quintic equation algorithm and the classic 1884 book by Klein[2] describing the relationship between the icosahedron and the solution of the quintic equation. At that stage this work lay fallow for more than a century since the algorithm for roots of the general quintic equation appeared intractable before the era of computers. Many of the key ideas appear to have been forgotten by the subsequent generations of mathematicians during the past century so that some of the underlying mathematics has the status of a lost art.

The work discussed in this book arose when I discovered the 1878 Kiepert paper and wished to see if the quintic equation algorithm described therein would work on a modern computer. I enlisted the help of a computer scientist, Prof. E. R. Canfield of the University of Georgia, to write a program that would use the Kiepert algorithm to solve the quintic equation (i.e., to calculate the roots of the quintic equation from its coefficients). This proved to be much more difficult than anticipated because of errors and lack of detail in the Kiepert paper. However, the quintic equation algorithm was finally made to work on an IBM-compatible personal computer but only after studying a considerable amount of related mathematics in several languages mainly in nineteenth century journals. This work led to the discovery of some unanticipated features of the quintic equation algorithm.

This book presents for the first time a complete algorithm for the roots of the general quintic equation with enough background information to make the key ideas accessible to nonspecialists and even to mathematically oriented readers who are not professional mathematicians. Two relatively short papers[3,4], have been published on this work on the quintic equation algorithm but the whole story is far too long to fit into a journal article of reasonable length. The book includes initial introductory chapters on group theory and symmetry, the Galois theory of equations, and some elementary properties of elliptic functions and associated theta functions in an attempt to make some of the key ideas accessible to less sophisticated readers. The book also includes discussion of the much simpler algorithms for roots of the general quadratic, cubic, and quartic equations before discussing the algorithm for the roots of the general quintic equation. The book concludes with a brief discussion of attempts to extend these ideas to algorithms for the roots of general equations of degrees higher than five.

I am indebted to Prof. E. R. Canfield of the Computer Science Department at the University of Georgia for helpful discussions during the varous stages of this project. In addition, I would like to acknowledge the patience and cooperation of my wife, Jane, during the preparation of this book.

R. Bruce King
Athens, Georgia
March, 1996

Bibliography

[1] L. Kiepert, Auflösung der Gleichungen Fünften Grades, *J. für Math.* **87** (1978), 114–133.

[2] F. Klein, *Vorlesungen über das Ikosaeder*, Teubner, Leipzig, 1884.

[3] R. B. King and E. R. Canfield, An Algebraic Algorithm for Calculating the Roots of a General Quintic Equation from its Coefficients, *J. Math. Phys.* **32** (1991), 823–825.

[4] R. B. King and E. R. Canfield, Icosahedral Symmetry and the Quintic Equation, *Computers and Mathematics with Applications* **24** (1992), 13–28.

Chapter 1

Introduction

Consider an algebraic equation of degree n, i.e.,

$$f(x) = a_0x^n + a_1x^{n-1} + \ldots a_n = \sum_{i=1}^{n} a_{n-i}x^i = 0 \qquad (1-1)$$

Algorithms for *solving* such equations, i.e., determining their roots, $x_1,\ldots,x_n$, as functions of their coefficients, a_n, have been of interest since ancient times.[1,2] Thus the solution of quadratic equations, $n = 2$, has been known since Babylonian times and the "quadratic formula"

$$x_n = \frac{-a_1 \pm \sqrt{a_1^2 - 4a_0a_2}}{2a_0} \qquad (1-2)$$

is familiar to high school students. A more complicated formula for solution of general cubic equations (equation 1–1 for $n = 3$), involving both square and cube roots, was found in the 16th century by Cardan, who built on previous work by Tartaglia and dal Ferro. Almost immediately thereafter Ferrari showed how the algorithms for solution of the quadratic and cubic equations could be combined to provide an algorithm for the solution of the general quartic equation (equation 1–1 for $n = 4$). *Thus all equations of degree 4 and lower can be solved by using only square and cube roots.* Solutions of algebraic equations (1–1) using functions no more complicated than radicals are called radical solutions or algebraic solutions of equations.

The success in finding radical solutions for all equations of degree 4 or lower naturally stimulated a search for radical solutions of the general quintic equation (equation 1–1 for $n = 5$). Naively it might appear that fifth roots

---

[1] M. Kline, *Mathematical Thought from Ancient to Modern Times*, Oxford Univ. Press, New York, 1972, pp. 263–272, 597–606, 752–763.
[2] J.-P. Tignol, *Galois' Theory of Algebraic Equations*, Longman, Essex, England, 1988.

would be sufficient for solution of any quintic equation just as square and cube roots are sufficient for solution of all equations of degree 4 and lower. Indeed the special *binomial* quintic equation of the type

$$x^5 - a_5 = 0 \qquad (1\text{--}3)$$

is soluble using fifth roots, i.e., $x_k = \sqrt[5]{a_5}$, just like any other binomial equation of degree m, i.e., $x^m - a_m = 0$, can be solved using the corresponding mth root, i.e., $x_k = \sqrt[m]{a_m}$. However, efforts by mathematicians such as Leibniz, Tschirnhausen, Euler, Vandermonde, Lagrange, and Ruffini to find a radical solution of the *general* quintic equation (equation 1–1 for $n = 5$ with $a_k \neq 0$ for $0 \leq k \leq 5$) all proved to be unsuccessful. Abel (1802–1829) then succeeded in proving the impossibility of solving by radicals the general equation of degree higher than four.[3] Galois (1811–1832) subsequently developed a theory, still known as *Galois theory*, which provided a method for characterizing equations that are solvable by radicals.[4] Galois' methods required the development of group theory to study the effect of permutations of the roots of the equation on functions of the roots. Thus the solution of algebraic equations was the first application of group theory, an area of algebra that has subsequently found many other applications in physical and mathematical sciences including the symmetry of physical objects. For example, the solution of the general quartic equation (equation 1–1 for $n = 4$) can be related to the 24 (= 4!) elements of symmetry in the tetrahedron (Figure 1–1a), and the solution of the general quintic equation (equation 1–1 for $n = 5$) can be related to the 120 (= 5!) symmetry elements in the icosahedron (Figure 1–1b).

---

[3]N. H. Abel, Beweis der Unmöglichkeit algebraische Gleichungen von höheren Graden als dem Viertem allgemein aufzulösen, *J. für Math.*, **1**, 65–84 (1826) = *Œuvres*, 1, 66–94.

[4]É. Galois, *Œuvres Mathématiques*, Gauthiers-Villars, Paris, 1897, pp. 33–50.

Introduction

Tetrahedron   Icosahedron

Figure 1–1: The tetrahedron and the icosahedron.

The insufficiency of radicals for solution of the general quintic equation naturally raises the question as to what type of functions are needed for its solution. The simplest type of quintic equation not soluble by radicals has the so-called Bring-Jerrard form

$$x^5 - a_4 x + a_5 = 0 \qquad (1-4).$$

The functions required to solve equations of this type as well as more complicated equations not solvable by radicals might be viewed as more complicated type of radicals, i.e., a *hyperradicals*, so that the development of the theory of such nonradical algebraic numbers might be viewed as an interesting branch of algebra. In this connection Hermite[5,6] first showed that elliptic modular functions provided solutions to the general quintic equation but the methods still appeared rather intractable. These ideas were developed further by Gordan[7] and Kiepert[8] in attempts to present complete algorithms for the solution of the general quintic equation. In a classic book Klein[9] presents in

---

[5] C. Hermite, Sur la Résolution de l'Équation du Cinquième Degré, *Compt. Rend.*, **46**, 508–515 (1858).

[6] C. Hermite, Sur l'Équation du Cinquième Degré, in C. Hermite, *Œuvres*, Volume II, pp. 347–424, Gauthier-Villars, Paris, 1908.

[7] P. Gordan, Über die Auflösung der Gleichungen vom Fünften Grade, *Math. Ann.*, **13**, 375-404 (1878).

[8] L. Kiepert, Auflösung der Gleichungen Fünften Grades, *J. für Math.*, **87**, 114–133 (1878).

[9] F. Klein, *Vorlesungen über das Ikosaeder*, Teubner, Leipzig, 1884.

detail the relationship between icosahedral symmetry and algebraic methods for solution of the general quintic equation. Although Klein's book[9] is an impressive synthesis of diverse important areas of 19th century mathematics, it does not provide information on the properties of the elliptic modular functions necessary to solve the general quintic equation.

The author has long been interested in applications of symmetry to chemical problems including icosahedral chemical structures.[10] This led to a natural curiosity about the relationship of icosahedral symmetry to the general quintic equation and then about the types of functions needed to solve the general quintic equation. However, none of the books on elliptic functions available in the extensive University of Georgia science library provided any information on the use of these functions to solve the general quintic equation. The classic paper by Kiepert[8] provides an algorithm on its last two pages for solution of the general quintic equation, but efforts to transfer this algorithm directly to a computer program were unsuccessful. Only after a detailed study of numerous 19th century mathematics papers in several languages as well as old algebra texts has it proven possible to assemble a working algorithm on a modern microcomputer for solution of the general quintic equation. This book is an effort to present the essential aspects of this classical and apparently largely forgotten mathematics in a modern form.

Since the concepts of group theory and symmetry are so central to the solution of algebraic equations, these concepts are first discussed in Chapter 2 with several geometric illustrations. Chapter 3 then extends the concept of symmetry to permutations of the roots of algebraic equations leading to the basic ideas of Galois theory. Other key concepts relating to solution of algebraic equations are also introduced in Chapter 3, notably the use of Tschirnhausen transformations to simplify the form of algebraic equations. Chapter 4 discusses the essential properties of elliptic functions and related integrals required for the solution of general algebraic equations of degree 5 or greater.

---

[10]R. B. King, The Icosahedron in Inorganic Chemistry, *Inorg. Chim. Acta*, **198–200**, 841–861 (1992).

Introduction

The remainder of the book applies these ideas to the solution of algebraic equations. Chapter 5 presents the classical methods for solutions of general equations of degrees no greater than 4 where radical methods can be applied as well as methods for identification of special quintic equations also soluble by radicals. Aspects of these methods pertaining to the structure of the symmetric groups S_n are given particular attention. Chapter 6 presents the details of the Kiepert algorithm[8] for solution of the general quintic equation using theta series. This algorithm has been verified on a microcomputer.[11,12] Chapter 7 summarizes the earlier methods of Hermite[5,6] and Gordan[7] for solving the general quintic equation. The final chapter, Chapter 8, goes beyond the quintic equation to summarize what is known about methods for solution of the general sextic equation, certain special septic equations, and the Umemura formula based on higher order theta functions for solution of the general algebraic equation of any degree.

---

[11] R. B. King and E. R. Canfield, An Algorithm for Calculating the Roots of a General Quintic Equation from its Coefficients, *J. Math. Phys.*, **32**, 823–825 (1991).
[12] R. B. King and E. R. Canfield, Icosahedral Symmetry and the Quintic Equation, *Comput. and Math. with Appl.*, **24**, 13–28 (1992).

Chapter 2

Group Theory and Symmetry

2.1 The Concept of Groups

A *group* is a collection or *set* of *elements* together with an inner binary operation, conventionally called *multiplication*, satisfying some special rules discussed below. It is not necessary to specify what the elements are in order to discuss the group that they constitute. Of particular interest in the context of this book are groups formed by permutations of sets of small numbers of objects, particularly the roots of algebraic equations; such groups are called *permutation groups*. Frequently the properties of such abstract permutation groups are most readily visualized by considering analogous or *isomorphic* groups formed by sets of symmetry operations on polyhedra or other readily visualized spatial objects. Groups formed by symmetry operations on three-dimensional objects are called *symmetry point groups*. Some features of both of these types of groups will be discussed in this chapter.

A set of elements combined with an binary operation (multiplication) to form a mathematical group must satisfy the following four conditions or rules[1]:

(1) The product of any two elements in the group and the square of each element must be an element of the group. A *product* of two group elements, AB, is obtained by applying the binary operation (multiplication) to them and the *square* of a group element, A^2, is the product of an element with itself. This definition can be extended to higher powers of group elements. The multiplication of two group elements is said to be *commutative* if the order of multiplication is immaterial, i.e., if $AB = BA$. In such a case A is said to *commute* with B. The multiplication of two group elements is *not* necessarily commutative.

(2) One element in the group must commute with all others and leave them unchanged. This element is conventionally called the *identity*

---

[1] F. J. Budden, *The Fascination of Groups*, Cambridge Univ. Press, London, 1972.

element and often designated as E. This condition may be concisely stated as $EX = XE = X$.

(3) The associative law of multiplication must hold. This condition may be expressed concisely as $A(BC) = (AB)C$, i.e., the result must be the same if C is multiplied by B to give BC followed by multiplication of BC by A to give $A(BC)$ or if B is multiplied by A to give AB followed by multiplication of AB by C to give $(AB)C$.

(4) Every element must have an inverse, which is also an element in the group. The element Z is the *inverse* of the element A if $AZ = ZA = E$. The inverse of an element A is frequently designated by A^{-1}. Note that multiplication of an element by its inverse is always commutative.

These defining characteristics of a group have been summarized concisely[2] by defining a group as "...a mathematical system consisting of elements with *inverses* which can be combined by some operation without going outside the system."

The number of elements in a group is called the *order* of the group. This book will be concerned exclusively with *finite groups*, i.e., groups with finite numbers of elements. Within a given group it may be possible to select various smaller sets of elements, each set including the identity element E, which are themselves groups. Such smaller sets are called *subgroups*. A subgroup of a group G is thus defined as a subset H of the group G which is itself a group under the same multiplication operation of G. The fact that H is a subgroup of G may be written $H \subset G$. The order of a subgroup must be an integral factor of the order of the group. Thus if H is a subgroup of G and $|H|$ and $|G|$ are the orders of H and G, respectively, then the quotient $|G|/|H|$ must be an integer. This quotient is called the *index* of the subgroup H in G.

Let A and X be two elements in a group. Then $X^{-1}AX = B$ will be equal to some element in the group. The element B is called the *similarity transform* of A by X and A and B may be said to be *conjugate*. Conjugate elements have the following properties:

---

[2] A. W. Bell and T. J. Fletcher, *Symmetry Groups*, Associated Teachers of Mathematics, 1964.

(1) **Every element is conjugate with itself.** Thus for any particular element A there must be at least one element X such that $A = X^{-1}AX$.

(2) **If A is conjugate with B, then B is conjugate with A.** Thus if $A = X^{-1}BX$, then there must be some element, Y, in the group such that $B = Y^{-1}AY$.

(3) **If A is conjugate with B and C, then B and C are conjugate with each other.**

A complete set of elements of a group which are conjugate to one another is called a *class* (or more specifically a *conjugacy class*) of the group. The number of elements in a conjugacy class is called its *order*; the orders of all conjugacy classes must be integral factors of the order of the group.

An important property of each element of a group is its *period*. In this context the period of an element is the minimum number of times that the element must be multiplied by itself before the identity element E is obtained, i.e., the smallest positive integer n such that $A^n = E$. The period of the identity element E is, of course, 1. The period of an element is sometimes also called its *order* but this is confusing because the term "order" is also used to describe the number of elements in a group or conjugacy class (see above).

Certain elements $g_1, g_2, \ldots, g_m$ of a given finite group G are called a set of *generators* if every element of G can be expressed as a finite product of their powers (including negative powers).[3] A set of generators may be denoted by the symbol $\{g_1, g_2, \ldots g_m\}$. A set of relations satisfied by the generators of a group is called an *abstract definition* or *presentation* of the group if every relation satisfied by the generators is an algebraic consequence of these particular relations. A group with only one generator (i.e., $m = 1$) is a *cyclic group*, $\{g\} \equiv C_n$, whose order n is the period of the single generator g, i.e., $g^n = E$, where E is the identity element. The cyclic group C_1 is the trivial group consisting solely of the identity element.

A group in which every element commutes with every other element is called a *commutative* group or an *Abelian* group after the famous Norwegian mathematician, Abel (1802–1829). In an Abelian group every element is in a

---

[3]H. S. M. Coxeter and W. O. J. Moser, *Generators and Relations for Discrete Groups*, Springer-Verlag, Berlin, 1972.

conjugacy class by itself, i.e., all conjugacy classes are of order one. A *normal subgroup* N of G, written $N \triangleleft G$, is a subgroup which consists only of *entire* conjugacy classes of G.[4] A *normal chain* of a group G is a sequence of normal subgroups $C_1 \triangleleft N_{a_1} \triangleleft N_{a_2} \triangleleft N_{a_3} \triangleleft \cdots \triangleleft N_{a_s} \triangleleft G$, in which s is the number of normal subgroups (besides C_1 and G) in the normal chain. A *simple* group has no normal subgroups other than the identity group C_1. Simple groups are particularly important in the theory of finite groups.[5,6] If a normal chain starts with the identity group C_1 and leads to G and if all of the quotient groups $N_{a_1}/C_1 = C_{a_1}, N_{a_2}/N_{a_1} = C_{a_2}, \ldots, G/N_{a_s} = C_{a_{s+1}}$ are cyclic, then G is a *composite* or *soluble* group.

2.2 Symmetry Groups

Many of the properties of groups are most readily visualized by studying the groups consisting of symmetry operations of polyhedra or other concrete three-dimensional objects. In this context a *symmetry operation* is a movement of an object such that, after completion of the movement, every point of the body coincides with an equivalent point or the same point of the object in its original orientation. The position and orientation of an object before and after carrying out a symmetry operation are indistinguishable. Thus a symmetry operation takes an object into an equivalent configuration.

The symmetry operations for objects in ordinary three-dimensional space can be classified into four fundamental types each of which is defined by a *symmetry element* around which the symmetry operation takes place. The four fundamental types of symmetry operations and their corresponding symmetry elements are listed in Table 2–1.

The identity operation, designated as E, leaves the object unchanged. Although this operation may seem trivial, it is mathematically necessary in order

---

[4] J. K. G. Watson, On the Symmetry Groups of Non-Rigid Molecules, *Mol. Phys.*, **21**, 577 (1971).
[5] D. Gorenstein, *Finite Groups*, Harper and Row, New York, 1968.
[6] D. Gorenstein, *Finite Simple Groups: An Introduction to their Classification*, Plenum, New York, 1982.

to convey the mathematical properties of a group on the set of all of the symmetry operations applicable to a given object. The reflection operation, designated as σ, involves reflection of the object through a plane, known as a *reflection plane*. For example, in a reflection through the xy-plane (conveniently designated as σ_{xy}) the coordinates of a point (x, y, z) change to $(x, y, -z)$—a reflection operation thus can result in the change of only a single coordinate. A rotation operation, designated as C_n, consists of a 360°/n rotation around a line, known as a *rotation axis*. For example, a C_2 rotation around the z-axis changes the coordinates of a point (x, y, z) to $(-x, -y, z)$—a rotation operation thus can result in a change of only two coordinates. An improper rotation, designated as S_n, consists of a 360°/n rotation around a line followed by a reflection in a plane perpendicular to the rotation axis. An S_2 operation is called an *inversion* and is designated by i; the intersection of the C_2-axis and the perpendicular reflection plane is called an *inversion center*. Inversion through the origin changes the coordinates of a point (x, y, z) to $(-x, -y, -z)$—thus an S_n operation must change the signs of all three coordinates. An S_1 improper rotation in which the C_1 proper rotation component is equivalent to the identity E corresponds to a reflection operation σ. Thus the reflection operation σ is a special type of improper rotation, namely S_1.

Table 2–1: The Four Fundamental Types of Symmetry Operations

Symmetry Operation	Designation	Corresponding Symmetry Element	Dimensions
Identity (no change)	E	The entire object	3
Reflection	σ	Reflection plane	2
Rotation	C_n	Rotation axis	1
Improper rotation	S_n	Improper rotation axis*	0

*Point of intersection of a proper rotation axis and a perpendicular reflection plane

Consider the set of symmetry operations in ordinary three-dimensional space describing the symmetry of an actual object. Such a set of symmetry operations satisfies the properties of a *group* in the mathematical sense and is therefore called a *symmetry point group*. In most cases such a symmetry point group contains a finite number of operations and is therefore a *finite group*.

There is a systematic way to classify objects in three-dimensional space by their symmetry point groups based on the following sequence of questions[7]:

(1) Is the object linear (i.e., only "one-dimensional")?

Linear objects are the only objects having infinite rather than finite symmetry point groups since the linear axis corresponds to an infinite order rotation axis, namely C_∞. If there is a reflection plane perpendicular to the infinite order rotation axis (dividing the object into two equivalent halves), then the symmetry point group is D_∞, if not the symmetry point group is C_∞.

(2) Does the object have multiple "higher-order" rotation axes (i.e., $C_{>2}$ axes)?

Multiple axes in this question refer to noncollinear $C_{>2}$ axes. Such nonlinear objects have the symmetries of well-known regular polyhedra and are generally readily recognizable by containing the regular polyhedra in some manner. Such groups are sometimes called the *polyhedral* point groups. Thus the tetrahedral groups T, T_h, and T_d with 12, 24, and 24 operations, respectively, have four noncollinear C_3 axes and are distinguished by the presence or absence of horizontal (σ_h) or diagonal (σ_d) reflection planes. The octahedral groups O and O_h with 24 and 48 operations, respectively, have not only four noncollinear C_3 axes but also three noncollinear C_4 axes and are distinguished by the presence or absence of reflection planes. Figure 2–1 shows the multiple rotation axes of the regular octahedron and the cube, both of which have the octahedral point group O_h. The icosahedral groups I and I_h with 60 and 120 operations, respectively, have ten noncollinear C_3 axes and six noncollinear C_5 axes and are also distinguished by the presence or absence of reflection planes.

---

[7] F. A. Cotton, *Chemical Applications of Group Theory*, Third Edition, Wiley, New York, 1990, Chapter 3.

3. If the object does not belong to either a linear group or a polyhedral group, then does it have proper or improper axes of rotation (i.e. C_n or S_n)?

If no axes of either type are found, the group is either C_s, C_i, or C_1 with 2, 2, and 1 operations, respectively, depending whether the object has a plane of symmetry (σ), and inversion center (i), or neither. The C_1 designation corresponds to an object with no symmetry at all and therefore to a symmetry point group containing only one element, namely the identity element E.

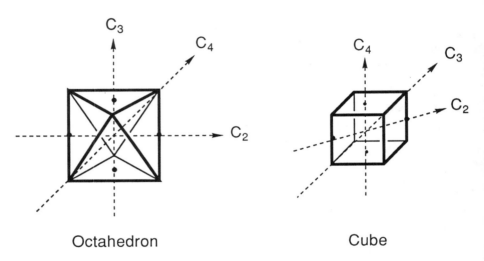

Octahedron   Cube

Figure 2–1: The multiple rotation axes in the octahedron and cube, both of which have O_h symmetry. In the octahedron the C_4 axes pass through opposite pairs of vertices, the C_3 axes pass through the midpoints (·) of opposite pairs of faces, and the C_2 axes pass through opposite pairs of edges. In the cube, the C_3 axes pass through opposite pairs of vertices, the C_4 axes pass through midpoints (·) of opposite pairs of faces, and the C_2 axes pass through opposite pairs of edges.

---

(4) Does the object have an *even*-order improper rotation axis S_{2n} but no planes of symmetry or any proper rotation axis other than one collinear with the improper rotation axis?

The presence of an even-order improper rotation axis S_{2n} without any noncollinear proper rotation axes or any reflection planes indicates the symmetry point group S_{2n} with $2n$ operations.

(5) **If the object does not belong to the linear point groups, the polyhedral point groups, or the point groups C_s, C_n, C_1, or S_{2n}, then look for the highest order rotation axis.** Call the highest order rotation axis C_n.

(6) **Are there n C_2 axes lying in a plane perpendicular to the C_n axis?** If there are n C_2 axes lying in a plane perpendicular to the C_n axis then the object belongs to one of the symmetry point groups D_n, D_{nh}, or D_{nd} with $2n$, $4n$, and $4n$ operations, respectively, depending on whether there are no planes of symmetry, a horizontal plane of symmetry (σ_h), or n vertical planes of symmetry (σ_v), respectively. If there are no C_2 axes lying in a plane perpendicular to the C_n axis, then the object belongs to one of the symmetry point groups C_n, C_{nv}, C_{nh}, with n, $2n$, and $2n$ operations, respectively, depending upon whether there are no planes of symmetry, n vertical planes of symmetry (σ_v), or a horizontal plane of symmetry (σ_h), respectively. If there are only C_2 axes, then a unique C_2 axis is chosen as the "reference axis" if there is any ambiguity as to which C_2 axis to choose.

2.3 Regular Polyhedra

The symmetries of regular polyhedra are important in the theory of the solution of algebraic equations. The symmetries of the five regular Platonic solids, namely the tetrahedron, octahedron, cube, icosahedron, and dodecahedron (Figures 2–1 and 2–2) were already recognized by the ancient Greeks. The fact that there are only five regular geometric solids in contrast to an infinite number of polygons must have been a major revelation to the ancient thinkers. The reason for only five regular geometric solids can be readily seen by the following arguments.[8,9]

---

[8]H. S. M. Coxeter, *Regular Polytopes*, Pitman Publishing Corp., New York, 1948.

 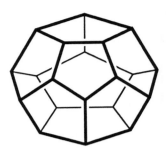

Tetrahedron **Icosahedron** **Dodecahedron**

Figure 2–2: The regular tetrahedron, icosahedron, and dodecahedron. The regular octahedron and cube are shown in Figure 2–1.

In order to have the solid angles required for forming closed polyhedra, a minimum of three edges must meet at a vertex. In this connection the number of edges meeting at a vertex is called the *degree* of the vertex. In the case of completely regular polyhedra the configuration of edges meeting at each vertex must be identical. If three equal-length edges meet at every vertex and if they join each other in equilateral triangular faces, the result is a tetrahedron. If instead, four or five equilateral triangles meet at each vertex, the result is an octahedron or an icosahedron, respectively. All of these regular polyhedra with equilateral triangular faces may be considered to have their vertices on finite surfaces with positive curvature such as the sphere. If six equilateral triangles meet at a vertex, an infinite flat two-dimensional lattice is obtained with a familiar hexagon structure with its vertices on an infinite surface with zero curvature. Such planar structures are called *tessellations* (Figure 2–3). More than six equilateral triangles meet at a vertex only with puckering and thus cannot give a regular solid. The vertices of such configurations can only be placed on infinite surfaces with negative curvature. Thus there are only three regular polyhedra with (equilateral) triangular faces, namely the tetrahedron,

---

[9] O. T. Benfey and L. Fikes, The Chemical Prehistory of the Tetrahedron, Octahedron, Icosahedron, and Hexagon, *Adv. Chem. Ser.*, **61**, 111–128 (1966).

octahedron, and icosahedron. Such polyhedra with triangular faces are called *deltahedra* since their faces are shaped like the Greek letter *delta*, Δ.

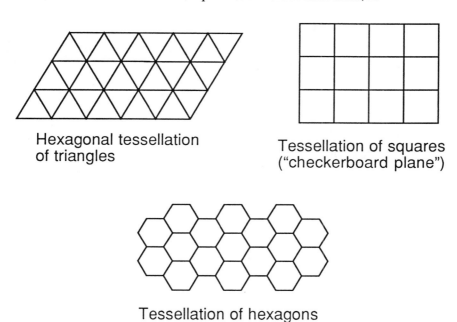

Figure 2–3: Examples of tessellations on a flat plane.

Additional regular solids can be formed from square or regular pentagonal faces. Vertices where three squares meet generate a cube whereas vertices where four squares meet generate a checkerboard plane (Figure 2–3). Vertices where three regular pentagons meet generate a regular dodecahedron; regular pentagons cannot form a tessellation. Regular hexagons with 120° angles can only form a tessellation and regular polygons with more than six sides have angles larger than 120° and thus cannot meet with two other equivalent polygons to form a solid angle at all.

The properties of the regular polyhedra are summarized in Table 2–2.

The regular polyhedra provide examples of pairs of dual polyhedra.[10] In this connection a given polyhedron P can be converted into its dual $P*$ by locating the centers of the faces of $P*$ at the vertices of P and the vertices of $P*$ above the centers of the faces of P. Two vertices in the dual $P*$ are connected by an edge when the corresponding faces in P share an edge. The process of dualization has the following properties:

(1) The numbers of vertices and edges in a pair of dual polyhedra P and $P*$ satisfy the relationships $v* = f$, $e* = e$, $f* = v$.

(2) Dual polyhedra have the same symmetry elements and thus belong to the same symmetry point group.

(3) Dualization of the dual of a polyhedron leads to the original polyhedron.

(4) The degree of a vertex in a polyhedron corresponds to the number of edges in the corresponding face in its dual.

Examples of pairs of dual polyhedra among the regular polyhedra are the octahedron/cube dual pair with O_h symmetry (Figure 2–4) and the icosahedron/dodecahedron dual pair with I_h symmetry. The tetrahedron with T_d symmetry is self-dual (i.e., dual to itself). For example, a degree 3 vertex of a polyhedron corresponds to a triangular face in its dual.

Table 2–2: Properties of the Regular Polyhedra

Polyhedron	Face Type	Vertex Degrees	Number of Edges	Number of Faces	Number of Vertices
Tetrahedron	Triangle	3	6	4	4
Octahedron	Triangle	4	12	8	6
Cube	Square	3	12	6	8
Dodecahedron	Pentagon	3	30	12	20
Icosahedron	Triangle	5	30	20	12

---

[10]B. Grünbaum, *Convex Polytopes*, Interscience, London, 1967, pp. 46–48.

Group Theory and Symmetry 17

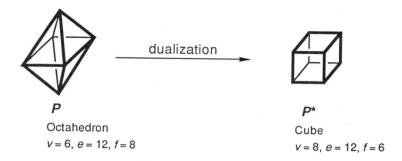

P   P*

Octahedron   Cube

$v = 6, e = 12, f = 8$ $v = 8, e = 12, f = 6$

Figure 2–4: The process of dualization to convert an octahedron into a cube.

The regular icosahedron and its dual, the regular dodecahedron, have the I_h symmetry point group with 120 elements. Its pure rotation subgroup, I, with 60 elements is isomorphic with the alternating group A_5. The symmetry point group of the regular tetrahedron, T_d, is a subgroup of I_h of index 5 but *not* a normal subgroup. Nevertheless, the ability to partition an object of icosahedral symmetry into five equivalent objects of tetrahedral or octahedral symmetry ($O_h \approx T_d \times C_2$) is an essential part of the Kiepert algorithm for solution of the general quintic equation (Chapter 6).

There are several ways of visualizing the partition of an object of icosahedral symmetry into five equivalent objects of at least tetrahedral symmetry thereby separating the effect of the fivefold axis of the icosahedron from that of its twofold and threefold symmetry elements. The 20 vertices of the dual of the icosahedron, namely the regular dodecahedron, can be partitioned into five sets of four vertices, each corresponding to a regular tetrahedron. Using the same idea Klein[11] partitions the 30 edges of an icosahedron into five sets of six edges each by the following method (Figure 2–5):

(1) A straight line is drawn from the midpoint of each edge through the center of the icosahedron to the midpoint of the opposite edge;

(2) The resulting 15 straight lines are divided into five sets of three mutually perpendicular straight lines.

---

[11] F. Klein, *Vorlesungen über das Ikosaeder*, Teubner, Leipzig, 1884, Part I, Chapter I, § 8.

Each of these five sets of three mutually perpendicular straight lines resembles a set of Cartesian coordinates and defines a regular octahedron. The construction is the dual of a construction depicted by Du Val[12] in color in which a regular dodecahedron is partitioned into five equivalent cubes. Alternatively, the 30 edge midpoints of an icosahedron can be used to form an icosidodecahedron having 30 vertices, 60 edges, 32 faces, and retaining icosahedral symmetry. The 30 vertices of this icosidodecahedron can then be partitioned into five sets of six vertices, each corresponding to a regular octahedron (Figure 2–6).

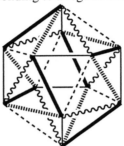

Figure 2–5: A regular icosahedron with its 30 edges partitioned into five sets of six edges each so that the midpoints of the edges in each set form a regular octahedron. The five sets of six edges each are indicated by the following types of lines: ——— ▬▬▬ ⅠⅠⅠⅠⅠ ∿∿∿ - - - -

Figure 2–6: The icosidodecahedron formed from the 30 edge-midpoints of a regular icosahedron and the partitioning of its 30 vertices into five sets of six vertices, each corresponding to regular octahedra. The six vertices in one of these sets are indicated by spades (♠).

---

[12]P. Du Val, *Homographies, Quaternions, and Rotations*, pp. 27–30 and Figs. 12–15, Oxford University Press, London, 1964.

2.4 Permutation Groups

The previous sections apply the concepts of group theory to symmetry point groups such as those describing the symmetry of readily visualized polyhedra. The concepts of group theory, of course, can be applied to more abstract sets such as the permutations of a set X of n objects, which, for example, may be the n roots of an algebraic equation of degree n. A set of permutations of n objects (including the identity permutation) with the structure of a group is called a *permutation group* of *degree n*.[13] Let G be a permutation group acting on the set X. Let g be any operation in G and x be any object in set X. The subset of X obtained by the action of all operations in G on x is called the *orbit* of x. The operations in G leaving x fixed is called the *stabilizer* of x; it is a subgroup of G and may be abbreviated as G_x. A *transitive* permutation group has only one orbit containing all objects of the set X. Sites permuted by a transitive permutation group are thus equivalent. Transitive permutation groups represent permutation groups of the highest symmetry and thus play a special role in permutation group theory.

The maximum number of distinct permutations of n objects is $n!$. The corresponding group is called the *symmetric* group of degree n and is traditionally designated as S_n (not to be confused with the designation S_n for an improper rotation of order n in Section 2.2). The symmetric group S_n is obviously the highest symmetry permutation group of degree n. All permutation groups of degree n must be a subgroup of the corresponding symmetric group S_n.

Let us now consider the structure of permutation groups. In this connection a permutation P_n of n objects can be described by a $2 \times n$ matrix of the following general type where the top row represents site labels and the bottom row represents object labels:

$$P_n = \begin{pmatrix} 1 & 2 & 3 & \dots & n \\ p_1 & p_2 & p_3 & \dots & p_n \end{pmatrix} \qquad (2.4\text{--}1)$$

---

[13]N. L. Biggs, *Finite Groups of Automorphisms*, Cambridge University Press, London, 1971.

The numbers $p_1, p_2, p_3, \ldots, p_n$ can be taken to run through the integers 1, 2, 3,...,n in some sequence. For a given n there are $n!$ possible different $\boldsymbol{P_n}$ matrices. The matrix $\boldsymbol{P_n}^0$ in which the bottom row $p_1, p_2, p_3, \ldots, p_n$ has the integers in the natural order 1, 2, 3,...,n (i.e., the bottom row of $\boldsymbol{P_n}^0$ is identical to the top row) can be taken to represent a reference configuration corresponding to the identity element in the corresponding permutation group.

Permutations can be classified as *odd* or *even* permutations based on how many *pairs* of numbers in the bottom row of the matrix $\boldsymbol{P_n}$ are out of their natural order. Alternatively, if the interchange of a single pair of numbers is called a *transposition*, the parity of a permutation corresponds to the parity of the number of transpositions. Thus a permutation which is obtained by an odd number of transpositions from the reference configuration is called an *odd* permutation and a permutation which is obtained by an even number of transpositions from the reference configuration is called an *even* permutation. The identity permutation corresponding to the reference configuration has zero transpositions and is therefore an even permutation by this definition.

A group can be defined relating the $\boldsymbol{P_n}$ matrices for a given n. First redefine the rows of $\boldsymbol{P_n}$ so that the top row represents the reference configuration $\boldsymbol{P_n}^0$ and the bottom row represents the object labels in any of the $n!$ possible permutations of the n objects. These permutations form a group of order $n!$ with the permutation leaving the reference configuration unchanged (i.e., that represented by $\boldsymbol{P_n}^0$ as so redefined) corresponding to the identity element, E. This permutation group is the symmetric group, S_n, of order $n!$ as defined above.

Now consider the nature of the operations in a symmetric permutation group S_n. These operations are permutations of labels which can be written as a product of cycles which operate on mutually exclusive sets of labels, e.g.

$$\begin{pmatrix} 1 & 2 & 3 & 4 & 5 & 6 \\ 2 & 4 & 5 & 1 & 3 & 6 \end{pmatrix} = (1\ 2\ 4)(3\ 5)(6) \qquad (2.4\text{--}2)$$

The cycle structure of a given permutation in the symmetric group S_n can be represented by a sequence of indexed variables, i. e., $x_1x_2x_3$ for the

Group Theory and Symmetry   21

permutation in equation 2.4–2. A characteristic feature of the symmetric permutation group S_n for all n is that all permutations having the same cycle structure come from the same conjugacy class.[14] Furthermore, no two permutations with different cycle structures can belong to the same conjugacy class. Therefore, for the symmetric permutation group S_n (but not necessarily for any of its subgroups) the cycle structures of permutations are sufficient to define their conjugacy classes. Furthermore, the number of conjugacy classes of the symmetric group S_n corresponds to the number of different partitions of n where a *partition of n* is defined as a set of positive integers $i_1, i_2, ..., i_k$ whose sum is n (equation 2.4–3).

$$\sum_{j=1}^{k} i_j = n \qquad (2.4\text{--}3)$$

An alternative presentation of conjugacy class information for the symmetric groups S_n is given by their cycle indices.[15,16,17] A *cycle index* $Z(S_n)$ for a symmetric permutation group S_n is a polynomial of the following form:

$$Z(S_n) = \sum_{i=1}^{i=c} a_i x_1^{c_{i1}} x_2^{c_{i2}} ... x_n^{c_{in}} \qquad (2.4\text{--}4)$$

In equation 2.4–4 c = number of conjugacy classes (i.e., partitions of n by equation 2.4–3), a_i = number of operations of S_n in conjugacy class i, x_j = dummy variable referring to cycles of length j, and c_{ij} = exponent indicating the number of cycles of length j in class i. These parameters in the cycle indices of the symmetric groups S_n must satisfy the following relationships:

(1)   Each of the $n!$ permutations of S_n must be in some class, i.e.,

---

[14] C. D. H. Chisholm, *Group Theoretical Techniques in Quantum Chemistry*, Academic Press, New York, 1976, Chapter 6.
[15] G. Pólya, Kombinatorische Anzahlbestimmungen für Gruppen, Graphen, und Chemische Verbindungen, *Acta Math.*, **68**, 145 (1937).
[16] G. Pólya and R. C. Reed, *Combinatorial Enumeration of Groups, Graphs, and Chemical Compounds*, Springer Verlag, New York, 1987.
[17] N. G. Debruin in *Applied Combinatorial Mathematics*, E. F. Beckenbach, Ed., Wiley, New York, 1964, Chapter 5.

$$\sum_{i=1}^{i=c} a_i = n! \qquad (2.4\text{--}5)$$

(2)   Each of the n objects being permuted must be in some cycle of each permutation in S_n (counting, of course, fixed points of cycles of length 1 represented by $x_1{}^{c_1}$), i.e.,

$$\sum_{j=1}^{j=n} jc_{ij} = n \text{ for } 1 \le i \le c \qquad (2.4\text{--}6)$$

The relationship between symmetry point groups and permutation groups as well as some important concepts can be illustrated by the symmetry point groups of the tetrahedron and trigonal bipyramid and the corresponding permutation groups on their vertices (Figure 2.7). The permutation group of the tetrahedron vertices is a transitive group since all four vertices are equivalent. However, the permutation group of the trigonal bipyramid vertices is an intransitive group partitioning its five vertices into two orbits, namely the two axial vertices (A in Figure 2.7) and the three equatorial vertices (E in Figure 2.7).

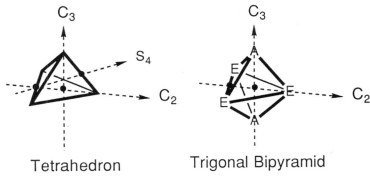

Tetrahedron   Trigonal Bipyramid

Figure 2.7:   The tetrahedron and the trigonal bipyramid showing the proper and improper rotation axes (Note that S_4 refers to an improper rotation axis of order 4 rather than the symmetric permutation group on four objects). The axial (A) and equatorial vertices (E) of the trigonal bipyramid are also shown.

Polyhedral polynomials are frequently expressed in terms of *homogeneous variables* so that all terms are the same combined degree in two variables, i.e.,

$$\sum_{i=1}^{n} a_{n-i} x^i = \sum_{i=1}^{n} a_{n-i} u^i v^{n-i} \qquad (2.5\text{–}1)$$

The use of homogeneous variables corresponds to the substitution $x = u/v$. The special symmetry of the regular polyhedra corresponds to the vanishing of some special functions of these polyhedral polynomials known as *transvectants* (Section 2.6).

Now consider the projection of points on the Riemann sphere onto its equatorial plane, taken as an Argand plane as noted above (Figure 2–8). A complex number $z = a + bi$ gives

$$a = \frac{p}{1-r}, \quad b = \frac{q}{1-r}, \quad a + bi = \frac{p + iq}{1-r} \qquad (2.5\text{–}2)$$

Solving for p, q, and r gives

$$p = \frac{2a}{1 + a^2 + b^2}, \quad q = \frac{2b}{1 + a^2 + b^2}, \quad r = \frac{-1 + a^2 + b^2}{1 + a^2 + b^2} \qquad (2.5\text{–}3)$$

Every rotation of the Riemann sphere around its center corresponds to a linear substitution

$$z' = \frac{\alpha z + \beta}{\gamma z + \delta} \qquad (2.5\text{–}4)$$

For a rotation of the sphere by an angle θ where p, q, r and $-p, -q, -r$ remain constant,

$$z' = \frac{(v + iu)z - (t - is)}{(t + is)z + (v - iu)} \qquad (2.5\text{–}5)$$

where $s = p \sin(\theta/2)$, $t = q \sin(\theta/2)$, $u = r \sin(\theta/2)$, and $v = \cos(\theta/2)$ so that

$$s^2 + t^2 + u^2 + v^2 = 1 \qquad (2.5\text{–}6)$$

For a rotation about the polar axis this reduces to

$$z' = e^{i\theta}z \tag{2.5-7}$$

Now consider the vertices of a regular octahedron and a regular icosahedron as points on the surface of such a Riemann sphere oriented such that the north pole ($z = \infty$) is one of the vertices in each case. This leads to the following homogeneous polynomials for these regular polyhedra in these orientations where z is taken to be u/v.[20]

(a) Octahedron (O_h symmetry):

$$\text{Vertices: } \tau = uv(u^4 - v^4) \tag{2.5-8a}$$

$$\text{Edges: } \chi = u^{12} - 33u^8v^4 - 33u^4v^8 + v^{12} \tag{2.5-8b}$$

$$\text{Faces: } W = u^8 + 14u^4v^4 + v^8 \tag{2.5-8c}$$

(b) Icosahedron (I_h symmetry):

$$\text{Vertices: } f = uv(u^{10} + 11u^5v^5 - v^{10}) \tag{2.5-9a}$$

$$\text{Edges: } T = u^{30} + 522u^{25}v^5 - 10{,}005\,u^{20}v^{10} - 10{,}005\,u^{10}v^{20}$$
$$- 522\,u^5v^{25} + v^{30} \tag{2.5-9b}$$

$$\text{Faces: } H = -u^{20} + 228\,u^{15}v^5 - 494\,u^{10}v^{10} - 228\,u^5v^{15} - v^{20}$$
$$\tag{2.5-9c}$$

The roots of these polyhedral polynomials correspond to the locations of the vertices, edge midpoints, and face midpoints on the Riemann sphere. Their degrees are equal to the numbers of corresponding elements (vertices, edges, or faces).

The following features are of interest concerning the polyhedral polynomials such as those in equations 2.5-8 and 2.5-9:

(1)   The vertices of a regular polygon with n vertices in the equatorial plane with one vertex at $z = 1$ correspond to the n roots of unity.

(2)   Derivation of the vertex polynomial of the octahedron, $\tau(u,v)$, is particularly obvious. Orientation of an octahedron so that two of the vertices are at the poles ($z = 0, \infty$) and one vertex is at $z = 1$ (Figure 2–9) forces the other three vertices to be at $z = -1, i,$ and $-i$ so that

$$\tau = z(z-1)(z+1)(z-i)(z+i) = z(z^4-1) = uv(u^4-v^4) \tag{2.5-10}$$

after converting to homogeneous variables by $z = u/v$.

---

[20]L. E. Dickson, *Modern Algebraic Theories*, Sanborn, Chicago, 1930, Chapter 13.

Group Theory and Symmetry

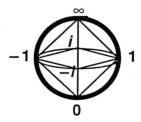

Figure 2–9: Orientation of the regular octahedron inside the Riemann sphere.

---

(3) The vertex and face polynomials of a polyhedron correspond to the face and vertex polynomials, respectively, of its dual. Thus the vertex polynomial τ (equation 2.5–8a) and face polynomial W (equation 2.5–8c) of the octahedron corresponds to the face polynomial and vertex polynomial, respectively, of its dual, namely the cube.

(4) A cube can be composed of two mutually dual tetrahedra, whose edges correspond to face diagonals of the cube (Figure 2–10). A single tetrahedron in this orientation has the following polynomial functions:

Vertices: $\Phi = u^4 + 2\sqrt{-3}\, u^2v^2 + v^4$ (2.5–11a)

Edges: $\tau = uv(u^4 - v^4)$ (2.5–11b)

Faces: $\Psi = u^4 - 2\sqrt{-3}\, u^2v^2 + v^4$ (2.5–11c)

Note that the tetrahedron edge polynomial (equation 2.5–11b) is the same as the octahedron vertex polynomial (equation 2.5–8a) since the midpoints of the six edges of the tetrahedron in this orientation are in the same locations ($z = 0, \infty, \pm 1, \pm i$) as the six vertices of the standard octahedron. Furthermore, the face polynomial Ψ of the tetrahedron is the same as the vertex polynomial of its dual. Since the four vertices of the tetrahedron and the four vertices of its dual together make the eight vertices of a cube, the product of the vertex functions of the two tetrahedra equals the vertex function of the cube made by the two tetrahedra $\Phi\Psi = W$.

Figure 2–10: Decomposition of a cube into two tetrahedra, whose vertices are indicated by **A** and **B**. The edges of the **A** tetrahedron are shown to the extent that they are visible.

---

(5) The special symmetries of the regular deltahedra lead to the following identities:

$$\text{Tetrahedron: } 12\sqrt{-3}\,\tau^2 - \Phi^3 + \Psi^3 \equiv 0 \quad \text{(degree 12)} \quad (2.5\text{–}12\text{a})$$
$$\text{Octahedron: } 108\,\tau^4 - W^3 + \chi^2 \equiv 0 \quad \text{(degree 24)} \quad (2.5\text{–}12\text{b})$$
$$\text{Icosahedron: } 1728\,f^5 - H^3 - T^2 \equiv 0 \quad \text{(degree 60)} \quad (2.5\text{–}12\text{c})$$

Note that the degrees of the left side of the identities (2.5–12) in (u,v) correspond to the orders of the pure rotation groups of the corresponding deltahedra, which are isomorphic to the permutation groups relevant to the solution of algebraic equations, i.e., $O \approx S_4$ for the quartic equation and $I \approx A_5$ for solution of the quintic equation. In this connection the icosahedral identity (equation 2.5–12c) will be seen to be important for the solution of the quintic equation.

2.6 Transvectants of Polyhedral Polynomials

The polyhedral polynomials in homogeneous form can be related by their transvectants derived from their partial derivatives by means of invariant theory.[21,22] In this connection the nth transvectant of two homogeneous polynomials $f(x,y)$ and $g(x,y)$, designated as $(f,g)^n$ is defined by the equation

---

[21] J. H. Grace and A. Young, *The Algebra of Invariants*, Cambridge, 1903.
[22] O. E. Glenn, *A Treatise on the Theory of Invariants*, Ginn and Co., Boston, 1915.

Group Theory and Symmetry 31

$$(f,g)^n = \sum_{k=1}^{n}(-1)^k \left(\frac{n!}{k!(n-k)!}\right)\left(\frac{\partial^n f(x,y)}{\partial x^{n-k}\partial y^k}\right)\left(\frac{\partial^n g(x,y)}{\partial x^k \partial y^{n-k}}\right) \qquad (2.6\text{–}1)$$

Of particular interest are the transvectants of a polynomial with itself, i.e., $(f,f)^n$.

Odd transvectants of the type $(f,f)^n$ (n = 1, 3, 5, 7,...) vanish since they have *even* numbers of terms of alternating sign which cancel out completely corresponding to the $(-1)^k$ factor in equation (2.6–1). The first transvectant $(f,g)^1$ is also known as the *functional determinant* of f and g, since it can be expressed by the determinant

$$(f,g)^1 = \begin{vmatrix} \frac{\partial f}{\partial x} & \frac{\partial f}{\partial y} \\ \frac{\partial g}{\partial x} & \frac{\partial g}{\partial y} \end{vmatrix} \qquad (2.6\text{–}2).$$

The second transvectants $(f,g)^2$ are also known as *Hessians* and can also be expressed by the determinant

$$(f,g)^2 = \begin{vmatrix} \frac{\partial^2 f}{\partial x^2} & \frac{\partial^2 f}{\partial x \partial y} \\ \frac{\partial^2 g}{\partial y \partial x} & \frac{\partial^2 g}{\partial y^2} \end{vmatrix} \qquad (2.6\text{–}3).$$

Let v, e, and f, be the number of vertices, edges, and faces of regular deltahedra, namely the tetrahedron, octahedron, and icosahedron. The functional determinants and Hessians relate the vertex ($\mathcal{V}(u,v)$ of degree v), edge ($\mathcal{E}(u,v)$ of degree e), and face ($\mathcal{F}(u,v)$ of degree f) functions by the following relationships where $k_{v,v}$ and $k_{v,f}$ are integers:

$$(\mathcal{V},\mathcal{V})^2 = k_{v,v}\,\mathcal{F} \qquad (2.6\text{–}4a)$$
$$(\mathcal{V},\mathcal{F})^1 = k_{v,f}\,\mathcal{E} \qquad (2.6\text{–}4b)$$

The fact that each differentiation step lowers the degree of the polynomial by 1 can be used to relate equation (2.6–4b) to Euler's theorem[23] as follows:

$$\deg(\mathcal{V},\mathcal{F})^1 = (v-1) + (f-1) = \deg(\mathcal{E}) = e$$

$\Rightarrow v + f - 2 = e$ corresponding to Euler's theorem. (2.6–5)

Similarly from equation (2.6–4a)

$$\deg(\mathcal{V},\mathcal{V})^2 = 2(v-2) = 2v - 4 = f. \quad (2.6\text{–}6).$$

For a deltahedron in which all faces are triangles

$$3f = 2e \Rightarrow e = {}^3\!/_2 f \quad (2.6\text{–}7)$$

so that Euler's theorem becomes

$$v + f - 2 = {}^3\!/_2 f \Rightarrow 2v - 4 = f \quad (2.6\text{–}8)$$

Note the resemblance of equations (2.6–6) and (2.6–8).

A special feature of the vertex polynomials for regular deltahedra, namely τ of equation (2.5–8a) for the octahedron, f of equation (2.5–9a) for the icosahedron, and Φ of equation (2.5–11a) for the tetrahedron is the identical vanishing of their fourth transvectants, i.e., $(\tau,\tau)^4 \equiv 0$, $(f,f)^4 \equiv 0$, and $(\Phi,\Phi)^4 \equiv 0$, respectively. This is a special indication of the symmetry of the regular deltahedra.

Figure 2–11 illustrates the procedure for calculating the second and fourth transvectants $(\tau,\tau)^2$ and $(\tau,\tau)^4$ of the octahedral vertex function $\tau = uv(u^4-v^4)$. Note that the Hessian $(\tau,\tau)^2$ is a multiple of the corresponding face function, namely $-25W$ whereas the fourth transvectant $(\tau,\tau)^4$ vanishes identically.

---

[23] B. Grünbaum, *Convex Polytopes*, Interscience Publishers, New York, New York, 1967, pp 130–142.

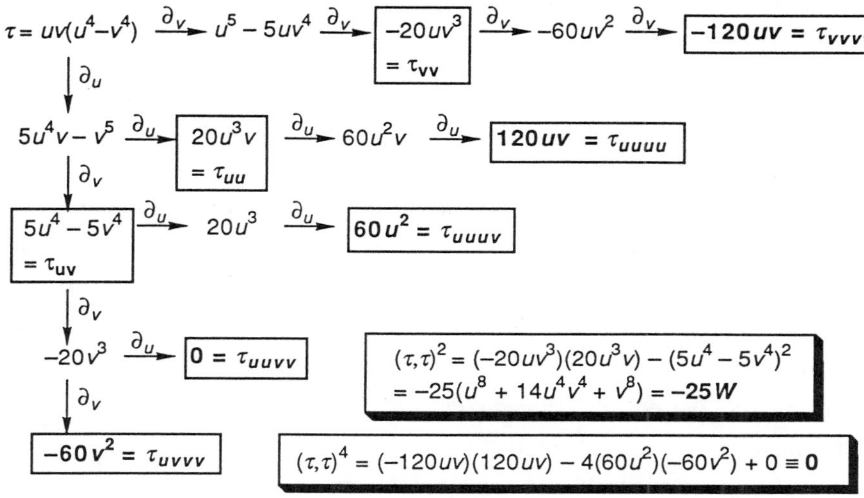

Group Theory and Symmetry

$$\tau = uv(u^4-v^4) \xrightarrow{\partial_v} u^5 - 5uv^4 \xrightarrow{\partial_v} \boxed{-20uv^3} \xrightarrow{\partial_v} -60uv^2 \xrightarrow{\partial_v} \boxed{-120uv = \tau_{vvvv}}$$

$$\downarrow \partial_u \qquad\qquad\qquad\qquad = \tau_{vv}$$

$$5u^4v - v^5 \xrightarrow{\partial_u} \boxed{20u^3v} \xrightarrow{\partial_u} 60u^2v \xrightarrow{\partial_u} \boxed{120uv = \tau_{uuuu}}$$

$$\downarrow \partial_v \qquad\quad = \tau_{uu}$$

$$\boxed{5u^4 - 5v^4} \xrightarrow{\partial_u} 20u^3 \xrightarrow{\partial_u} \boxed{60u^2 = \tau_{uuuv}}$$
$$= \tau_{uv}$$

$$\downarrow \partial_v$$

$$-20v^3 \xrightarrow{\partial_u} \boxed{0 = \tau_{uuvv}} \qquad (\tau,\tau)^2 = (-20uv^3)(20u^3v) - (5u^4 - 5v^4)^2$$
$$= -25(u^8 + 14u^4v^4 + v^8) = -25W$$

$$\downarrow \partial_v$$

$$\boxed{-60v^2 = \tau_{uvvv}} \qquad (\tau,\tau)^4 = (-120uv)(120uv) - 4(60u^2)(-60v^2) + 0 \equiv 0$$

Figure 2–11: Illustration of the calculation of the second transvectant $(\tau,\tau)^2$ and the fourth transvectant $(\tau,\tau)^4$ of the octahedral vertex function $\tau = uv(u^4-v^4)$. Subscripts indicate differentiation variables and ∂_u and ∂_v indicate differentiation with respect to u and v, respectively.

Chapter 3

The Symmetry of Equations: Galois Theory and Tschirnhausen Transformations

3.1 Rings, Fields, and Polynomials

The previous chapter uses group theory to study spatial and permutational symmetry. This chapter extends the application of group theory to the symmetry inherent in algebraic equations, which is closely linked to the methods required for their solution. The group-theoretical aspects of algebraic equations were first introduced by Évariste Galois (1811–1832) so that this area of mathematics is frequently called *Galois theory*. An excellent discussion of Galois theory is given in a book by Stewart.[1]

The concepts of a ring, an integral domain, and a field are essential to Galois theory. Such concepts express in a rigorous way the properties of numbers of various types and allow a precise definition of the concept of a *polynomial* corresponding to the left side of the general algebraic equation

$$f(x) = a_0 x^n + a_1 x^{n-1} + \ldots a_n = \sum_{i=1}^{n} a_{n-i} x^i = 0 \qquad (3.1\text{--}1).$$

These concepts are defined as follows:

(1) Ring: A *ring* is defined as a set R equipped with two binary operations, namely addition (+) and multiplication (×), such that R is an Abelian group under addition, multiplication is associative, and the following distributive laws hold for all $a, b \in R$:

$$(a + b) \times c = (a \times c) + (b \times c) \qquad (3.1\text{--}2a)$$
$$a \times (b + c) = (a \times b) + (a \times c) \qquad (3.1\text{--}2b)$$

The operation of addition has an additive identity called 0 such that

$$a + 0 = a \qquad (3.1\text{--}3).$$

The operation of multiplication $a \times b$ may be abbreviated as ab. Note that a ring has *two* binary operations whereas a group (Section 2.1) has only a single binary operation.

---

[1] I. Stewart, *Galois Theory*, Chapman and Hall, London, 1973.

(2) Integral domain: An *integral domain* is a ring D with the following three additional properties:
  (a) Multiplication in D is commutative, i.e., $ab = ba$ for all a, b.
  (b) There exists an element $1 \in D$ such that $a1 = 1a = a$ for all $a \in D$;
  (c) If $ab = 0$ for $a, b \in D$ then either $a = 0$ or $b = 0$.

(3) Field: A *field* is a ring F such that $F \backslash \{0\}$ (i.e., the field without the additive identity) is an abelian group under multiplication. Thus every non-zero element $a \in F$ has a multiplicative inverse a^{-1} where ab^{-1} may also be written as division, i.e., a/b or $\dfrac{a}{b}$.

Frequently encountered examples of such structures include the integral domain **Z** of integers and the fields **Q**, **R**, and **C** of rational, real, and complex numbers, respectively.

The concepts of subring, subfield, and ideal are also useful. A *subring* of a ring R is a nonempty subset S such that if $a, b \in S$, then $a + b \in S$, $a - b \in S$, and $ab \in S$. A *subfield* of a field F is a subset S containing the elements 0 and 1 such that if $a, b \in S$, then $a + b \in S$, $a - b \in S$, $ab \in S$, and further if $a \neq 0$, then $a^{-1} \in S$. An *ideal* of a ring R is a subring I such that if $i \in I$ and $r \in R$ then ir and ri are in I. Thus **Z** is a subring of **Q**, and **R** is a subfield of **C** while the set $2\mathbf{Z}$ of even integers is an ideal of **Z**. If I is an ideal of the ring R, the quotient ring R/I consists of the cosets of I in R with the following operations where $r, s \in R$ and $I + r$ is the coset $\{i + r: i \in I\}$:

$$(I + r) + (I + s) = I + (r + s) \tag{3.1-4a}$$
$$(I + r)(I + s) = I + (rs) \tag{3.1-4b}$$

Thus the set of integers divisible by a fixed integer n, namely $n\mathbf{Z}$, is an ideal of **Z** and the quotient ring $\mathbf{Z}_n = \mathbf{Z}/n\mathbf{Z}$ is the *ring of integers modulo n (or* mod $n)$. The elements of the ring $\mathbf{Z}_n$ are conventionally written as $0, 1, 2,\ldots,n-1$.

The following property of $\mathbf{Z}_n$ is important:

Theorem 3.1–1: The ring $\mathbf{Z}_n$ is a field if and only if n is a prime number.
Proof: (a) Suppose n is not prime. If $n = 1$, then $\mathbf{Z}_n = \mathbf{Z}/\mathbf{Z}$ which has only one element and thus cannot be a field. If $n > 1$, then $n = rs$ where r and s are integers less than n. If $I = n\mathbf{Z}$, then $(I + r)(I + s) = I + rs = I$ by equations 3.1–4. But I is the zero element of $\mathbf{Z}/I$ while $I + r$ and $I = s$ are non-zero.

Since in a field the product of two non-zero elements is non-zero, $\mathbf{Z}/I$ cannot be a field.

(b) Suppose n is prime. Let $I + r$ be a non-zero element of $\mathbf{Z}/I$. Since r and n are coprime, there exist integers a and b such that $ar + bs = 1$. Then $(I + a)(I + r) = (I + 1) - (I + n)(I + b) = I + 1$ and $(I + r)(I + a) = I + 1$. Since $I + 1$ is the identity element of $\mathbf{Z}/I$, we have found a multiplicative inverse for the given element $I + r$. Thus every non-zero element of $\mathbf{Z}/I$ has an inverse, so that $\mathbf{Z}_n = \mathbf{Z}/I$ is a field. ∎

The *prime subfield* of a field K is defined to be the intersection of all subfields of K and is thus the unique smallest subfield of K. The rational numbers $\mathbf{Q}$ and the field $\mathbf{Z}_p$ (p prime) have no proper subfields and are thus equal to their prime subfields. It can be shown that these are the only fields which can occur as prime subfields in the following theorem:

Theorem 3.1–2: Every prime subfield is isomorphic either to the field $\mathbf{Q}$ of rational numbers or the field $\mathbf{Z}_p$ of integers modulo a prime number.

Proof: Let K be a field and P its prime subfield. P contains 0 and 1 and therefore contains the elements n^* ($n \in \mathbf{Z}$) defined by

$$n^* = 1 + 1 + \ldots + 1 \text{ (} n \text{ times) if } n > 0 \qquad (3.1\text{–}5a)$$

$$0^* = 0 \qquad (3.1\text{–}5b)$$

$$n^* = -(-n)^* \text{ if } n < 0 \qquad (3.1\text{–}5c)$$

Two distinct cases arise:

(a) $n^* = 0$ for some $n \neq 0$. Since also $(-n)^* = 0$, there exists a smallest positive integer p such that $p^* = 0$. If p is composite, say $p = rs$ where r and s are smaller positive integers, then $r^*s^* = p^* = 0$ so that either $r^* = 0$ or $s^* = 0$ contrary to the definition of p. Therefore p is prime. The elements n^* form a ring isomorphic to $\mathbf{Z}_p$, which is a field by Theorem 3.1–1. This must be the whole of P since P is the smallest subfield of K.

(b) $n^* \neq 0$ if $n \neq 0$. Then P must contain all of the elements m^*/n^* where m and n are integers and $n \neq 0$, These form a subfield isomorphic to $\mathbf{Q}$ by the map which sends m^*/n^* to m/n which is necessarily the whole of P. ∎

A field K with a prime subfield isomorphic to $\mathbf{Q}$ is said to have *characteristic* 0. A field K with a prime subfield isomorphic to $\mathbf{Z}_p$ is said to have *characteristic p*.

This leads to the following lemmas[1]:

The Symmetry of Equations

Lemma 3.1-3: If K is a subfield of L, then K and L have the same characteristic.

Lemma 3.1-4: If k is a non-zero element of the field K and if n is an integer such that $nk = 0$, then n is a multiple of the characteristic of K.

Sometimes it is possible to embed a ring R into a field, i.e., to find a field containing a subring isomorphic to R. For example, the integers **Z** can be embedded into the rational numbers **Q** in a way that every element of **Q** is a fraction whose numerator and denominator lie in **Z**. This concept is generalized by defining a *field of fractions* of the ring R as a field K containing a subring R'' isomorphic to R such that every element of K can be expressed in the form r/s for $r, s \in R''$ where $s \neq 0$. It can be shown that every integral domain possesses a field of fractions and that for any given integral domain R, all fields of fractions are isomorphic. An element $r \in R$ may be identified with its image $[r,1]$ in the corresponding field of fractions F so that $[r,s]$ is the fraction r/s in F.

These concepts allow a precise logical definition to be given of a *polynomial*, which is the left side of the general algebraic equation 3.1–1. Let R be a ring in which multiplication is commutative. A polynomial over R in the indeterminate x can be defined by the following expression where $r_0,\ldots,r_n \in R$, $0 \leq n \in$ **Z**, and x is undefined:

$$r_0 + r_1 x + \ldots + r_n x^n \tag{3.1-6}$$

The elements $r_0,\ldots r_n$ are called the *coefficients* of the polynomial. If $r_n = 1$, then the polynomial is called a *monic* polynomial. Terms of the type $0x^m$ and $1x^m$ are frequently omitted and written as x^m, respectively. Two polynomials are *defined* to be equal if and only if the corresponding coefficients of *all* terms are equal. The set of all polynomials over the ring R in the indeterminate x is also a ring $R[x]$ with the same operations as in R. The ring $R[x]$ is called the *ring of polynomials over* R *in the indeterminate x*. If R is an integral domain, $R[x]$ has a field of fractions as suggested by the following lemma:

Lemma 3.1-5: If R is an integral domain and x is an indeterminate, then $R[x]$ is an integral domain.

Proof: Suppose that $f = f_0 + f_1 x + \ldots + f_n x^n$ and $g = g_0 + g_1 x + \ldots + g_m x^m$ where $f_n \neq 0 \neq g_m$ and all of the coefficients f_k and g_k lie in R. The coefficient of x^{m+n} in fg is $f_n g_m$ which is non-zero since R is an integral domain. Thus if f

and g are non-zero, then fg is non-zero indicating that $R(x)$ is an integral domain. ∎

The field of fractions necessarily possessed by the integral domain $R[x]$ is called the *field of rational expressions in x over R* and can be denoted by $K(x)$, where K is a field of fractions of R.

If f is a polynomial over a commutative ring R and $f \neq 0$, then the degree of f, designated as ∂f, is the highest power of the indeterminate x occurring in f with a non-zero coefficient. If R is an integral domain and f and g are polynomials over R, then the degrees of the polynomials have the following relationships:

$$\partial(f+g) \leq \text{maximum }(\partial f, \partial g) \qquad (3.1\text{–}7a)$$

$$\partial(fg) = \partial f + \partial g \qquad (3.1\text{–}7b)$$

The $\leq$ rather than $=$ sign in equation 3.1–7a is necessary to allow for the special case of cancellation of the highest degree terms in f and g.

A polynomial may always be divided by another polynomial provided that a remainder term is allowed leading to the following theorem:

Theorem 3.1–6: If f and g are polynomials over a field K and $f \neq 0$, then there exist unique polynomials q and r such that $g = fq + r$ where r has smaller degree than f.

Proof: The proof uses induction on the degree of g. If $\partial g = -\infty$, then $g = 0$, and q and r may also be taken to be 0. If $\partial g = 0$, then $g = k \in K$. If also $\partial f = 0$, then $f = j \in K$, and we may take $q = k/j$ and $r = 0$. Otherwise $\partial f > 0$ so that we may take $q = 0$ and $r = g$ to start the induction.

Now assume that the theorem holds for all polynomials of degree $< n$ and let $\partial g = n > 0$. If $\partial f > \partial g$ we may as before take $q = 0$ and $r = g$. Otherwise we may take $f = a_m x^m + \ldots + a_0$ and $g = b_n x^n + \ldots + b_0$ where $a_m \neq 0 \neq b_n$ and $m \leq n$. Now start the division process g/f by putting $g_1 = b_n a_m^{-1} x^{n-m} f - g$ so that terms of the highest degree cancel, and $\partial g_1 < \partial g$. By induction we can find polynomials q_1 and r_1 such that $g_1 = fq_1 + r_1$ and $\partial r_1 < \partial f$. Now let $q = b_n a_m^{-1} x^{n-m} - q_1$ and $r = -r_1$ so that $g = fq + r$ and $\partial r < \partial f$ as required.

Finally we prove the uniqueness of $g = fq + r$. Suppose that $g = fq_1 + r_1 = fq_2 + r_2$ where $\partial r_1, \partial r_2 < \partial f$. Then $f(q_1 - q_2) = r_1 - r_2$. However

The Symmetry of Equations

$f(q_1 - q_2)$ must have a higher degree than $r_1 - r_2$ unless both are zero. Since $f \neq 0$, $q_1 = q_2$ and $r_1 = r_2 \Rightarrow q$ and r are unique. ∎

Using this notation q is called the *quotient* and r is called the *remainder* upon dividing g by f. The process of finding q and r is sometimes called the Division Algorithm or Euclidean Algorithm.

The next step is to introduce the concepts of divisibility and highest common factor for polynomials. Thus if f and g are polynomials over a field K, then we say that f divides g (or f is a factor of g or g is a multiple of f) if there exists some polynomial h over K such that $g = fh$. The notation $f|g$ means that f divides g whereas the notation $f \nmid g$ means that f does not divide g. The polynomial d over K is called a *highest common factor* (*hcf*) of f and g if $d|f$ and $d|g$, and further, whenever $e|f$ and $e|g$, then $e|d$.

The following theorems relate to the properties of highest common factors:

Theorem 3.1–7: If d is a hcf of the polynomials f and g over a field K and if $0 \neq k \in K$, then kd is also a hcf for f and g. If d and e are two hcfs for f and g, then there exists a non-zero element $k \in K$ such that $e = kd$.

Proof: Clearly $kd|f$ and $kd|g$. If $e|f$ and $e|g$, then $e|d$ so that $e|kd$. Therefore kd is a hcf. If d and e are hcfs, then by definition $e|d$ and $d|e$ so that $e = hd$ for some polynomial h. Since $e|d$ the $\partial e \leq \partial d$ so that $\partial h \leq 0 \Rightarrow h = k \in K$. Since $0 \neq e = kd$, then $k \neq 0$. ∎

Theorem 3.1–8: If f and g are non-zero polynomials over a field K and d is a hcf for f and g, then there exist polynomials a and b over K such that $d = af + bg$.

Proof: Let $f = r_{-1}$ and $g = r_0$. Dividing f by g leads to the successive remainders $r_{-1} = q_1 r_0 + r_1$ $(\partial r_1 \leq \partial r_0)$, $r_0 = q_2 r_1 + r_2$ $(\partial r_2 \leq \partial r_1)$, $r_1 = q_3 r_2 + r_3$ $(\partial r_3 \leq \partial r_2)$, ..., $r_i = q_{i+2} r_{i+1} + r_{i+2}$ $(\partial r_{i+2} \leq \partial r_{i+1})$, etc. Since the degrees of the r_i decrease, a point must eventually be reached where the process stops, when $r_{s+2} = 0$, so that the final equation is $r_s = q_{s+2} r_{s+1}$. Let $d = r_{s+1}$ since hcfs are unique up to constant factors.

The induction hypothesis for proof of Theorem 3.1–8 is that there exist polynomials a_i and b_i such that $d = a_i r_i + b_i r_{i+1}$. Using the above notation this is clearly true when $i = s + 1$ for then we may take $a_i = b_i = 0$. However, $r_{i+1} = r_{i-1} - q_{i+1} r_i$, and by induction $d = a_i r_i + b_i(r_{i-1} - q_{i+1} r_i)$ so that if $a_{i-1} = b_i$ and

$b_{i-1} = a_i - b_i q_{i+1}$, then $d = a_{i-1} r_{i-1} + b_{i-1} r_i$. By descending induction $d = a_{-1} r_{-1} + b_{-1} r_0 = af + bg$ where $a = a_{-1}$ and $b = b_{-1}$ thereby completing the proof.∎

These concepts allow the development of a theory of the factorization of polynomials. A polynomial over a commutative ring is said to be *reducible* if it is a product of two polynomials of smaller degree. Otherwise it is *irreducible*. Thus all polynomials of degree 0 or 1 are irreducible since they certainly cannot be expressed as a product of polynomials of smaller degree. The polynomial $x^2 - 2$ is irreducible over the rational numbers **Q** but can be reduced over the real numbers **R** by the equation

$$x^2 - 2 = (x - \sqrt{2})(x + \sqrt{2}) \tag{3.1-8}.$$

This leads to the following theorems relating to the factorization of polynomials:

Theorem 3.1–9: Any non-zero polynomial over a field K is a product of irreducible polynomials over K.

Proof: Let g be any non-zero polynomial over K and use induction on the degree of g. If $\partial g = 0$ or 1 then g is automatically irreducible. If $g \geq 1$, either g is irreducible or $g = hj$ where ∂h and $\partial j < \partial g$. By induction h and j are products of irreducible polynomials; therefore g is also such a product. The theorem follows by induction.∎

Theorem 3.1–10: For any field K, the factorization of polynomials over K into irreducible polynomials is unique up to constant factors and the order in which the factors are written.

Proof: Suppose that $f = f_1 \cdots f_r = g_1 \cdots g_s$ where f is a polynomial over K and $f_1, \ldots, f_r$ and $g_1, \ldots, g_s$ are irreducible polynomials over K. If all the f_i are constant then $f \in K$ so that all of the g_j are constant. Otherwise we may assume that no f_i is constant by dividing out all the constant terms. Then $f_1 | g_1 \cdots g_s$ and $f_i | g_i$ for some i. Choose the notation so that $i = 1$, and then $f_1 | g_1$. Since f_1 and g_1 are irreducible and f_1 is not a constant, we must have $f_1 = k_1 g_1$ for some constant k_1. Analogously $f_2 = k_2 g_2, \ldots, f_r = k_r g_r$ where $k_2, \ldots, k_r$ are constants. The remaining g_j ($j > r$) must also be constant or else the degree of the right hand side would be too large. The theorem follows.∎

Methods for testing the irreducibility of polynomials are very difficult just like methods for testing the primality of numbers. The following theorems are relevant to testing for the irreducibility of polynomials:

The Symmetry of Equations

Theorem 3.1–11: If f is a polynomial over the integers $\mathbf{Z}$ which is irreducible over $\mathbf{Z}$, then f, considered as a polynomial over the rationals $\mathbf{Q}$, is also irreducible over $\mathbf{Q}$ so that factorization over the integers $\mathbf{Z}$ is equivalent to factorization over the rationals $\mathbf{Q}$.

Proof: Suppose that f is irreducible over $\mathbf{Z}$ but reducible over $\mathbf{Q}$ so that $f = gh$ where g and h are polynomials over $\mathbf{Q}$ of smaller degree. Multiplying through by the products of the denominators of the coefficients of g and h gives $nf = g'h'$ where $n \in \mathbf{Z}$ and g' and h' are polynomials over $\mathbf{Z}$. We now show that we can cancel out all of the prime factors of n one by one without going outside $\mathbf{Z}[x]$.

Suppose that p is a prime factor of n, and let $g' = g_0 + g_1 x + \ldots + g_r x^r$ and $h' = h_0 + h_1 x + \ldots + h_s t^s$. Then either p divides all of the coefficients g_i or else p divides all of the coefficients h_j. If not, there must be *smallest* values i and j such that $p \nmid g_i$ and $p \nmid h_i$. However, p divides the coefficient of x^{i+j} in $g'h'$ which is $h_0 g_{i+j} + h_1 g_{i+j-1} + \ldots h_j g_i + \ldots + h_{i+j} g_0$, and by the choice of i and j the prime p divides every term of this expression except perhaps $h_j g_i$. But p divides the whole expression so that $p | h_j g_i$. This contradicts $p \nmid g_i$ and $p \nmid h_i$ thereby proving the theorem. ∎

Theorem 3.1–12 (Eisenstein's Irreducibility Criterion): Let $f(x) = a_0 + a_1 x + \ldots + a_n x^n$ be a polynomial over $\mathbf{Z}$. Suppose that there is a prime q such that $q \nmid a_n$, $q | a_i$ ($0 \leq i \leq n-1$), and $q^2 \nmid a_0$. Then f is irreducible over $\mathbf{Q}$.

Proof: By Theorem 3.1–11 it suffices to show that f is irreducible over $\mathbf{Z}$. Suppose for a contradiction $f = gh$ where $g = b_0 + b_1 x + \ldots + b_r x^r$, and $h = c_0 + c_1 x + \ldots + c_s x^s$ are polynomials of smaller degree over $\mathbf{Z}$. Then $r + s = n$. Now $b_0 c_0 = a_0$ so that $q | b_0$ or $q | c_0$. However, by assumption q cannot divide both b_0 and c_0 so that without loss of generality we can assume $q | b_0$ and $q \nmid c_0$. If all coefficients b_i were divisible by q, then a_n would be divisible by q contrary to assumption. Let b_i be the first coefficient of g not divisible by q so that $a_i = b_i c_0 + \ldots + b_0 c_i$ where $i < n$. This implies that q divides c_0 since q divides $a_i, b_0, \ldots, b_{i-1}$ but not b_i. This leads to a contradiction thereby showing that f is irreducible. ∎

If $f(x)$ is a polynomial over a commutative ring R, the *zeros* of $f(x)$ correspond to the values of $\alpha \in R$ where $f(\alpha) = 0$. The following theorem relates the zeros of $f(x)$ to its factors:

Theorem 3.1–13: If $f(x)$ is a polynomial over the field K, then an element $\alpha \in K$ is a zero of $f(x)$ if and only if $(x - \alpha)|f(x)$.

Proof: If $(x - \alpha)|f(x)$ then $f(x) = (x - \alpha)g(x)$ for some polynomial g over K so that $f(\alpha) = (\alpha - \alpha)g(\alpha) = 0$. Conversely, suppose $f(\alpha) = 0$. By Theorem 3.1–6 there exist polynomials q and r over K such that $f(x) = (x - \alpha)q(x) + r(x)$ where $\partial r < 1$. Thus $r(x) = r \in K$. Substituting α for x gives $0 = f(\alpha) = (\alpha - \alpha)q(\alpha) + r$ so that $r = 0$ and $(x - \alpha)|f(x)$. ∎

If f is a polynomial over the field K, an element $\alpha \in K$ is a *simple* zero of f if $(x - \alpha)|f(x)$ but $(x - \alpha)^2 \nmid f(x)$. Similarly the element $\alpha \in K$ is a zero of f of *multiplicity* m if $(x - \alpha)^m | f(x)$ but $(x - \alpha)^{m+1} \nmid f(x)$. For example, the zero at $\alpha = 1$ is a simple zero for $x^2 + x + 2 = (x - 1)(x + 2)$ but a zero of multiplicity 2 for $x^3 + 3x - 2 = (x - 1)^2(x + 2)$. The following theorems relate to the multiplicity of zeros of polynomials:

Theorem 3.1–14: If f is a non-zero polynomial over the field K with distinct zeros $\alpha_1, \ldots, \alpha_r$ with multiplicities $m_1, \ldots, m_r$, respectively, then $f(x) = (x - \alpha_1)^{m_1} \ldots (x - \alpha_r)^{m_r} g(x)$ where $g(x)$ has no zeros in K. Conversely if $f(x) = (x - \alpha_1)^{m_1} \ldots (x - \alpha_r)^{m_r} g(x)$ where $g(x)$ has no zeros in K, then the zeros of f in K are $\alpha_1, \ldots, \alpha_r$ with multiplicities $m_1, \ldots, m_r$, respectively.

Proof: For any $\alpha \in K$ the polynomial $x - \alpha$ is irreducible. Hence for any distinct α and $\beta \in K$ the polynomials $x - \alpha$ and $x - \beta$ are coprime. The uniqueness of factorization (Theorem 3.1–10) leads to the relation $f(x) = (x - \alpha_1)^{m_1} \ldots (x - \alpha_r)^{m_r} g(x)$, and g cannot have any zeros in K or else f would have extra zeros or zeros of larger multiplicity. The converse follows simply from the uniqueness of factorization (Theorem 3.1–10). ∎

Theorem 3.1–15: The number of zeros of a polynomial, counted according to multiplicity, is less than or equal to its degree.

Proof: This follows from Theorem 3.1–14 where $m_1 + \ldots + m_r \leq \partial f$. ∎

The original theory of Galois used polynomials over the complex field **C**. More recently in the 1920s and 1930s the methods of Galois theory were generalized to arbitrary fields. The central object of study then became the field extension related to a polynomial rather than the polynomial itself. Thus every polynomial f over a field K was used to generate another field L containing K (or a subfield isomorphic to K). In this connection, a *field extension* is defined as a map $i: K \to L$ where K and L are fields; K is the small field and L is the large

The Symmetry of Equations 43

field. The inclusion maps $i_1: \mathbf{Q} \to \mathbf{R}$, $i_2: \mathbf{R} \to \mathbf{C}$, and $i_3: \mathbf{Q} \to \mathbf{C}$ relating the rational, real, and complex numbers are all examples of field extensions.

If $i: K \to L$ is a field extension, K can usually be identified with its image $i(K)$ so that i can be considered an inclusion map and K a subfield of L. Under such circumstances the notation $L:K$ can also be used for field extensions where L is an extension of K. If K is a field and X is a nonempty subset of K, then the subfield of K *generated by* X is the intersection of all subfields of K which contain X.

Frequently in the case of a field extension $L:K$ we are interested in fields lying between K and L. In such cases we are interested in subsets X of L containing K, i.e., sets of the form $K \cup Y$ where $Y \subseteq L$. If $L:K$ is an extension and Y is a subset of L, then the subfield of L generated by $K \cup Y$ is written $K(Y)$ and said to be obtained from K by adjoining $Y \subseteq L$. In general $K(Y)$ is considerably larger than $K \cup Y$. Thus the subfield $\mathbf{R}(i)$ of $\mathbf{C}$ must contain all elements $x + iy$ where $x,y \in R$ so that $\mathbf{C} = \mathbf{R}(i)$. Furthermore, the subfield of $\mathbf{R}$ consisting of all elements $p + q\sqrt{2}$ where $p,q \in \mathbf{Q}$ is $\mathbf{Q}(\sqrt{2})$.

A field extension $L:K$ is a *simple extension* if it has the property $L = K(\alpha)$ for some $\alpha \in L$; the extensions $\mathbf{R}(i)$ and $\mathbf{Q}(\sqrt{2})$ are both examples of simple field extensions. In addition the field extension $\mathbf{Q}(\sqrt{2}, \sqrt{3})$ may be reduced to the simple extension $\mathbf{Q}(\sqrt{2} + \sqrt{3})$ because of the relationships $(\sqrt{2} + \sqrt{3})^2 = 5 + 2\sqrt{6}$ and $\sqrt{6}((\sqrt{2} + \sqrt{3})) = 2\sqrt{3} + 3\sqrt{2}$ so that $\mathbf{Q}(\sqrt{2}, \sqrt{3}) = \mathbf{Q}(\sqrt{2} + \sqrt{3})$. An *isomorphism* between two field extensions $i: K \to K^*$, $j: L \to L^*$ is a pair (λ, μ) of field isomorphisms $\lambda: K \to L$, $\mu: K^* \to L^*$ such that for all $k \in K$, $j(\lambda(k)) = \mu(i(k))$.

Simple extensions can be considered to be algebraic or transcendental depending upon their relationship to polynomials. Let $K(\alpha):K$ be a simple extension. If there exists a non-zero polynomial p over K such that $p(\alpha) = 0$, then α is an *algebraic* element over K and the extension $K(\alpha):K$ is a *simple algebraic extension*. If there is no non-zero polynomial p of any finite degree α where $p(\alpha) = 0$, then α is *transcendental* over K, and $K(\alpha):K$ is a *simple transcendental extension*. The following theorem gives a method for constructing a simple transcendental extension of the field K:

Theorem 3.1–16: The field of rational expressions $K(x)$ is a simple transcendental extension of the field K.

Proof: The extension $K(x)$ is clearly a simple extension. If f is a polynomial over K such that $f(x) = 0$, then $f = 0$ by definition of $K(x)$. ∎

If $L:K$ is a field extension and $\alpha \in L$ is algebraic over K, then the *minimum polynomial* of α over K is defined to be the unique monic polynomial m over K of smallest degree such that $m(\alpha) = 0$. The following theorems relating to minimum polynomials are relevant:

Theorem 3.1-17: If α is an algebraic element over the field K, then the minimum polynomial of α over K is irreducible over K and divides every polynomial of which α is a zero.

Proof: Suppose that the minimum polynomial m of α over K is reducible so that $m = fg$ where f and g are of smaller degree. Assume that f and g are monic. Since $m(\alpha) = 0$ by the definition of minimum polynomial, we have $f(\alpha)g(\alpha) = 0$ so that either $f(\alpha) = 0$ or $g(\alpha) = 0$. However, this contradicts the definition of m as the *minimum* polynomial leading to the conclusion that m must be irreducible over K. ∎

Theorem 3.1-18: Let $K(\alpha):K$ be a simple algebraic extension where α has a minimum polynomial m over K. Then any element of $K(\alpha)$ has a unique expression in the form $p(\alpha)$ where p is a polynomial over K and $\partial p < \partial m$.

Proof: Every element of $K(\alpha)$ can be expressed in the form $f(\alpha)/g(\alpha)$ where $f, g \in K[x]$ and $g(\alpha) \neq 0$ since the set of all such elements is a field, contains K and α, and lies inside $K(\alpha)$. Since $g(\alpha) \neq 0$, m does not divide g and since m is irreducible, m and g are coprime, i.e., their hcf is 1. By Theorem 3.1-8 there exist polynomials a and b over K such that $ag + bm = 1$. Hence $a(\alpha)/g(\alpha) = 1$, so that $f(\alpha)/g(\alpha) = f(\alpha)a(\alpha) = h(\alpha)$ for some polynomial h over K. Let r be the remainder on dividing h by m. Then $r(\alpha) = h(\alpha)$. Since $\partial r < \partial m$ the existence of such an expression is proved.

In order to show uniqueness suppose that $f(\alpha) = g(\alpha)$ where ∂f and $\partial g < \partial m$. If $e = f - g$ then $e(\alpha) = 0$ and $\partial e < \partial m$. By definition of m we have $e = 0$ so that $f = g$ thereby proving the theorem. ∎

In addition, it can be proved[1] that if K is any field and m is any irreducible monic polynomial over K, then there exists an extension $K(\alpha):K$ such that α has minimum polynomial m over K.

The degree of an field extension can be obtained by associating a vector space structure with the extension. Thus if $L:K$ is a field extension, the

The Symmetry of Equations

operations $(\lambda,u) \to \lambda u$ ($\lambda \in K$, $u \in L$) and $(u,v) \to u + v$ ($u,v \in L$) define on L the structure of a vector space over K.[1] The *degree* $[L:K]$ of a field extension $L:K$ is the dimension of L considered as a vector space over K by this procedure. For example, the field **C** of complex numbers is two-dimensional over the real numbers **R** with $\{1,i\}$ being a basis so that $[\mathbf{C}:\mathbf{R}] = 2$. It can be shown that if K, L, and M are fields and $K \subseteq L \subseteq M$, then $[M:K] = [M:L][L:K]$. The degree of a simple extension can easily be found by the following theorem[1]:

Theorem 3.1–19: Let $K(\alpha):K$ be a simple extension. If it is transcendental, then $[K(\alpha):K] = \infty$. If it is algebraic, then $[K(\alpha):K] = \partial m$ where m is the minimum polynomial of α over K.

Proof: For the transcendental case it suffices to note that the elements 1, α, α^2,... are linearly independent over K so that the corresponding vector space is infinite dimensional leading to infinite degree. For the algebraic case a basis is determined. Let $\partial m = n$ and consider the n elements 1, α, ..., α^{n-1}. By Theorem 3.1–18 these elements span $K(\alpha)$ over K, and by the uniqueness clause of Theorem 3.1–18 they are linearly independent. Therefore they form a basis, and $[K(\alpha):K] = n = \partial m$. ∎

A *finite extension* is one whose degree is finite. Any simple algebraic extension is thus finite. However, the converse is not true. In this connection an extension $L:K$ is *algebraic* if every element of L is algebraic over K. It can be shown[1] that $L:K$ is a finite extension if and only if L is algebraic over K and there exist finitely many elements $\alpha_1,...,\alpha_s \in L$ such that $L = K(\alpha_1,...,\alpha_s)$.

3.2 Galois Theory: Solubility of Algebraic Equations by Radicals

Group theory was invented by Galois to study the permutations of the zeros of polynomials. Thus any polynomial $f(x)$ has a group of permutations of its zeros, now called its *Galois group*, whose structure is closely related to the methods required for solving the corresponding polynomial equation $f(x) = 0$.

The problem of finding the Galois group of a given polynomial is clearly an important one. Let K be a subfield of the field L. An automorphism

α of L is defined to be a *K-automorphism* of L if $\alpha(k) = k$ for all $k \in K$. This effectively makes α an automorphism of the *extension* $L:K$ rather than just the large field L. The set of all K-automorphisms of L in the field extension $L:K$ is easily seen to form a group under composition; this group is called the *Galois group* $\Gamma(L:K)$. The role of Galois groups in solving polynomial equations relates to the observation originally by Galois that under certain extra hypotheses discussed later there is a one-to-one correspondence between subgroups of the Galois group of $L:K$ and subfields M of L such that $K \subseteq M$.

A very simple illustration of a Galois group uses the extension **C:R** and the complex conjugation operation. If α is an **R**-automorphism of **C** and $j = \alpha(i)$, then $j^2 = [\alpha(i)]^2 = \alpha(i^2) = \alpha(-1) = -1$ since $\alpha(r) = r$ for all $r \in$ **R**. Thus either $j = i$ or $j = -i$. Now $\alpha(x + iy) = \alpha(x) + \alpha(i)\alpha(y) = x + jy$ for any $x, y \in$ **R**. This leads to two candidates for **R**-automorphisms, namely $\alpha_1: x + iy \to x + iy$ (the identity operation) and $\alpha_2: x + iy \to x - iy$ (the complex conjugation operation). The maps α_1 and α_2 can be shown to be **R**-automorphisms. Since $\alpha_2^2 = \alpha_1$, the Galois group $\Gamma(\mathbf{C}:\mathbf{R})$ is the cyclic group of order 2 (C_2).

If $L:K$ is a field extension, a field M such that $K \subseteq M \subseteq L$ is called an *intermediate field*. A group $M^\spadesuit = \Gamma(L:M)$ of all M-automorphisms of L can be associated with each intermediate field M. Using this terminology, $K^\spadesuit$ is the entire Galois group, and $L^\spadesuit = 1$, namely the identity map on L. Thus if the field M is a subfield of N, i.e., $M \subseteq N$, then the group $N^\spadesuit$ is a subgroup of $M^\spadesuit$, i.e., $N^\spadesuit \subseteq M^\spadesuit$. Conversely, if each subgroup H of $\Gamma(L:K)$ has an associated set $H^\dagger$ of all elements $x \in L$ such that $\alpha(x) = x$ for all $\alpha \in H$, the set $H^\dagger$ can be shown to be a subfield of L containing K.[1] The field $H^\dagger$ is called the *fixed field* of the subgroup H.

The concept of a splitting field can now be introduced which relates to the phenomenon that a polynomial with no zeros over one field may have zeros over a larger field. Thus the polynomial $x^2 + 1$ has no zeros over **R** but the zeros $\pm i$ over **C**. Every polynomial can be resolved into a product of linear factors indicating its full complement of zeros if the original field is extended to a suitable splitting field. If K is a field and f is a polynomial over K, then f is said to *split* over K if f can be expressed as a product of linear factors $f(x)$

$k(x-\alpha_1)\ldots(x-\alpha_n)$ where $k,\alpha_1,\ldots,\alpha_n \in K$. The field Σ is defined as a *splitting field* for the polynomial f over the field K if the following conditions are met:
(1) $K \subseteq \Sigma$,
(2) f splits over Σ,
(3) f splitting over Σ' if $K \subseteq \Sigma' \subseteq \Sigma$ implies $\Sigma' = \Sigma$.

A splitting field can be constructed by adjoining to the field K some additional elements which are zeros of f in accord with the following theorem:

Theorem 3.2–1: If K is any field and f is any polynomial over K then there exists a splitting field for f over K.

Proof: Induction is used on the degree ∂f. If $\partial f = 1$ there is nothing to prove for f splits over K. If f does not split over K, then it has an irreducible factor f_1 of degree > 1. In this case adjoin α to K where $f(\alpha) = 0$. Then in $K(\alpha)[x]$ we have $f = (x - \alpha)g$ where $\partial g = \partial f - 1$. By induction there is a splitting field Σ for g over $K(\alpha)$. But then Σ is a splitting field for f over K.∎

Furthermore if $i: K \to K'$ is a field isomorphism, if T is a splitting field for f over K, and if T' is a splitting field for $i(f)$ over K', then the extensions $T:K$ and $T':K'$ can be shown to be isomorphic.

The concept of a splitting field can be illustrated by the equation
$$f(x) = (x^2 - 3)(x^3 + 1)$$
$$= (x + \sqrt{3})(x - \sqrt{3})(x + 1)(x - \frac{-1+i\sqrt{3}}{2})(x - \frac{-1-i\sqrt{3}}{2}) \quad (3.2-1).$$

The splitting field for this equation inside **C** can readily be shown to be $\mathbf{Q}(\sqrt{3}, i)$.

Normal field extensions are particularly important in Galois theory. In this connection an algebraic field extension $L:K$ is defined to be *normal* if every irreducible polynomial f over K which has at least one zero in L splits in L. For example, the algebraic field extension **C:R** is normal since every polynomial whether or not irreducible splits in **C**. An example of an algebraic field extension which is not normal is $\mathbf{Q}(\sqrt[3]{2})$ since the irreducible polynomial $x^3 - 2$ has a zero, namely $\sqrt[3]{2}$ in $\mathbf{Q}(\sqrt[3]{2})$, but does not split in $\mathbf{Q}(\sqrt[3]{2})$. A normal finite extension such as **C:R** has a well-behaved Galois group in the sense that the Galois correspondence is a bijection. However, the nonnormal extension $\mathbf{Q}(\sqrt[3]{2}):\mathbf{Q}$ has a badly behaved Galois group. The following theorem indicates the close connection between normal extensions and splitting fields:

Theorem 3.2–2: An extension $L:K$ is normal and finite if and only if L is a splitting field for some polynomial over K.
Proof: If $L:K$ is normal and finite, it can be expressed as $L = K(\alpha_1,\ldots,\alpha_s)$ for certain α_i algebraic over K. Let m_i be the minimum polynomial of α_i over K and let $f = m_1 \cdots m_s$. Each m_i is irreducible over K and has a zero $\alpha_i \in L$ so by normality each m_i splits over L. Therefore f splits over L. Since L is generated by K and the zeros of f, it is a splitting field for f over K. ■

The additional concept of separability is required for fields of non-zero characteristic. An irreducible polynomial f over a field K is *separable* over K if it has no multiple zeros in a splitting field. This means that in any splitting field f can be expressed as $f(x) = k(x - \sigma_1)\ldots(x - \sigma_n)$ where the σ_i are all different.

With these concepts in mind the fundamental properties of the Galois correspondence can be established between a field extension and its Galois group. Let $L:K$ be a finite separable normal algebraic field extension of degree n with Galois group G, which consists of all K-automorphisms of L, let $\mathcal{F}$ be the set of intermediate fields M, and let $\mathcal{G}$ be the set of all subgroups H of G. Furthermore, if $M \in \mathcal{F}$, then $M^{\clubsuit}$ is the group of all M-automorphisms of L, and if $H \in \mathcal{G}$, $H^{\dagger}$ is the fixed field of H. The fundamental theorem of Galois theory then states the following:

(1) The Galois group G has order n;

(2) The maps $\clubsuit$ and $\dagger$ are mutual inverses and establish an order-reversing one-to-one correspondence between $\mathcal{F}$ and $\mathcal{G}$;

(3) If M is an intermediate field, then the degree of the extension $L:M$ is equal to the order of the group $M^{\clubsuit}$, i.e., $[L:M] = |M^{\clubsuit}|$, and the degree of the extension $M:K$ is equal to $|G|/|M^{\clubsuit}|$;

(4) An intermediate field M is a normal extension of K if and only if $M^{\clubsuit}$ is a normal subgroup of G (Section 2.1);

(5) If an intermediate field M is a normal extension of K, then the Galois group of $M:K$ is isomorphic to the quotient group $G/M^{\clubsuit}$.

Details of the proof of this theorem are given elsewhere.[1]

The Galois correspondence expressed by the above theorem can be used to derive a condition which must be satisfied by an equation soluble by radicals, namely that its Galois group must be a soluble group (Section 2.1). In this connection an extension $L:K$ is defined to be a *radical extension* if $L =$

The Symmetry of Equations

$K(\alpha_1,\ldots,\alpha_m)$ where for each $i = 1,\ldots,m$, there exists an integer $n(i)$ such that $\alpha_i{}^{n(i)} \in K(\alpha_1,\ldots,\alpha_{n-1})$. The elements α_i are called a *radical sequence* for $L:K$. Informally, a radical extension is obtained by a sequence of adjunctions of nth roots for various values of n. If f is a polynomial over a field K of characteristic zero and Σ is a splitting field of f over K, then f is said to be *soluble by radicals* if there exists a field M containing Σ such that $M:K$ is a radical extension. The Galois group of f over K is the Galois group $\Gamma(\Sigma:K)$.

The Galois group G of a polynomial f over the field K with a splitting field Σ over K can be viewed as a group of permutations on the zeros of f. Thus if $\alpha \in \Sigma$ is a zero of f, then $f(\alpha) = 0$ so $f[g(\alpha)] = g[f(\alpha)] = 0$ for any $g \in G$. In this way every element $g \in G$ induces a permutation g' of the set of zeros of f in Σ thereby defining an injection of G into the permutation group of the zeros of f.

The irreducibility of a polynomial f over the field K (Section 3.1) can be related to the transitivity of its Galois group G. Consider a reducible polynomial $f = gh$ where g and h are irreducible so that $\partial f = \partial(gh) = \partial g + \partial h$ (see equation 3.1–7b). The roots of the algebraic equation $f(x) = g(x)h(x) = 0$ consist of the union of the ∂g roots of $g(x)$ and the ∂h roots of $h(x)$ so that determining the roots of $f(x)$ requires solving one equation of degree ∂g, namely $g(x)$, and one equation of degree ∂h, namely $h(x)$ rather than the more complicated task of solving a single equation of degree ∂f, namely $f(x)$. This, of course, is how factoring an algebraic equation simplifies determination of its roots. The Galois group G of $f(x)$ is an intransitive group with two orbits. One orbit corresponds to the ∂g zeros of $g(x)$, and the other orbit corresponds to the ∂h zeros of $h(x)$. In this way intransitive Galois groups can be seen to correspond to reducible polynomials and transitive Galois groups to irreducible polynomials.

Methods for solving irreducible algebraic equations are thus seen by Galois theory to correspond to transitive permutation groups, i.e., the Galois groups of the irreducible equations. The transitive permutation groups of degrees less than eight, and thus corresponding to irreducible equations of degrees lower than eight, are listed in Table 2–3. All of the transitive groups of degree 4 or lower are soluble groups so that all algebraic equations of degree 4 or lower are soluble by radicals (Sections 5.1 and 5.2). The lowest degree simple group is the degree 5 group A_5 of order $5!/2 = 60$, which means that the general quintic equation with A_5 or S_5 Galois group cannot be solved by radicals. However, quintic equations with the dihedral group D_5 of order 10 or the metacyclic group M_5 of order 20 can be solved by radicals since these groups are soluble (Section 5.3). Recent papers have given algorithms using radicals to determine the roots of such solvable quintic equations.[2,3]

The insolubility of quintic equations having the simple Galois group A_5 using only radicals does not mean that they cannot be solved using functions more complicated than radicals. Thus suppose that for any real number a the *ultraradical* $\sqrt[*]{a}$ is defined as the real zero of $x^5 + x - a$. Then Jerrard has shown that the general quintic equation can be solved by the use of such ultraradicals (Chapter 7),[4] which turn out to be elliptic functions (Chapter 4). However, conversion of the general quintic equation to the so-called Bring-Jerrard equation $x^5 + x - a$ requires a very difficult Tschirnhausen transformation (Section 3.3),[5] which is avoided in the alternative method of Kiepert[6] for solution of the general quintic equation (Chapter 6).

Table 2-3 containing the transitive permutation groups of degree less than eight suggests other types of equations with simple Galois groups preventing them from being solved using only radicals. All of the alternating groups A_n ($n \geq 5$) are simple which means that the *general* equations of any degree greater than five cannot be solved using only radicals. The difficulties of solving such equations appear to increase factorially according to the $n!/2$ order of the alternating group A_n. A complete solution of the general sextic equation, including all of the necessary Tschirnhausen transformations (Section 3.3), does not appear to have been worked out. However, a special sextic equation, theoretically obtainable from the general sextic equation by Tschirnhausen

---

[2] D. S. Dummit, Solving Solvable Quintics, *Math. Comp.*, **57**, 387 (1991).
[3] S. Kobayashi and H. Nakagawa, Resolution of Solvable Quintic Equations, *Math. Japonica*, **37**, 883 (1992).
[4] A. Hausner, The Bring-Jerrard Equation and Weierstrass Elliptic Functions, *Amer. Math. Monthly*, **69**, 193 (1962).
[5] A. Cayley, On Tschirnhausen's Transformation, *Phil. Trans. Roy. Soc. London*, **151**, 561 (1861); Cayley's Collected Works, paper 275.
[6] L. Kiepert, Auflösung der Gleichungen fünften Grades, *J. für Math.*, **87**, 114 (1878).

transformations, has been solved using double (genus 2) theta functions (Section 8.1). In the case of the septic (degree 7) equation, simple groups arise not only for the general septic equation with A_7 or S_7 Galois groups but also for the special septic equation with the simple L(3,2) Galois group of order 168 (Table 2.1).[7] In the 19th century Klein[8] and then Radford[9] studied this special septic equation in some detail (Section 8.2).

3.3 Tschirnhausen Transformations

Consider the general monic algebraic equation of degree n, i.e.,

$$x^n + a_1 x^{n-1} + \ldots a_n = x^n + \sum_{i=1}^{n-1} a_{n-i} x^i = 0 \qquad (3.3-1).$$

A useful prelude to the solution of such an equation may involve substituting a new variable

$$y_k = \alpha_0 + \alpha_1 x_k + \ldots + \alpha_{n-1} x_k^{n-1} = \sum_{j=0}^{n-1} \alpha_j x_k^j \qquad (3.3-2)$$

where x_k ($1 \leq k \leq n$) is a root of equation 3.3-1 to give a new monic algebraic equation of the same degree, i.e.,

$$y^n + A_1 y^{n-1} + \ldots + A_n = y^n + \sum_{i=1}^{n-1} A_{n-i} y^i = 0 \qquad (3.3-3).$$

Such a procedure is called a *Tschirnhausen transformation*[10,11] and is beneficial when the coefficients α_m ($0 \leq m \leq n-1$) in equation 3.3-2 are chosen so that equation 3.3-3 is simpler to solve than equation 3.3-1. After the transformed equation 3.3-3 is solved to give the roots y_k ($0 \leq k \leq n$), then the

---

[7] H. Weber, *Lehrbuch der Algebra*, Vieweg, Braunschweig, 1898, Volume 2, §§ 141–147.
[8] F. Klein, Über die Transformation siebenter Ordnung der elliptischen Funktionen, *Math. Ann.*, **14**, 428–471 (1879), *Gesammelte Mathematische Abhandlungen*, Volume 3, pp. 90–136, Springer Verlag, Berlin, 1923.
[9] E. M. Radford, On the Solution of Certain Equations of the Seventh Degree, *Quart. J. Math.*, **30**, 263–306 (1898).
[10] O. Perron, *Algebra*, Third Edition, de Gruyter, Berlin, 1951, Chapter 2, § 16.
[11] L. E. Dickson, *Modern Algebraic Theories*, Sanborn, Chicago, 1930, Chapter 12.

Tschirnhausen transformation is undone by solving equation 3.3–2 to give the roots x_k ($0 \leq k \leq n$) as functions of y_k. The value of Tschirnhausen transformations to facilitate the solution of algebraic equations thus relates not only to the ease of solving the transformed equation 3.3–3 but also to the degree of the equation 3.3–2 expressing y_k in terms of x_k.

The simplest example of a Tschirnhausen transformation is used in the solution of the monic cubic equation

$$x^3 + a_1 x^2 + a_2 x + a_3 = 0 \qquad (3.3\text{–}4).$$

A new variable y can be defined by the linear equation

$$y_k = x_k + \frac{a_1}{3} \qquad (3.3\text{–}5)$$

which corresponds to equation 3.3–2 with $\alpha_1 = a_1/3$, and $\alpha_2 = 0$. This gives the monic cubic equation

$$y^3 + 3A_2 y + A_3 = 0 \qquad (3.3\text{–}6)$$

in which the y^2 term vanishes, i.e., $A_1 = 0$. A cubic equation with the y^2 term missing can readily be solved by well-known methods using nested square and cube roots or by using trigonometric functions (Section 5.1). Solving the linear equation 3.3–5 to give x_k as a function of y_k is trivial and gives

$$x_k = y_k - \frac{a_1}{3} \qquad (3.3\text{–}7)$$

Thus after the roots y_k of equation 3.3–6 are found, the corresponding roots x_k of equation 3.3–4 can be found using equation 3.3–7.

The Tschirnhausen transformation is frequently useful for simplifying algebraic equations when the coefficients α_k of equation 3.3–2 are so chosen that some of the coefficients A_k of equation 3.3–3 vanish. This can be done by using the Newton relationships between the sums of the roots of equations 3.3–1 and 3.3–3 and their coefficients, i.e.,

$$\Sigma x_k + a_1 = 0 \qquad (3.3\text{–}8a)$$
$$\Sigma x_k^2 + a_1 \Sigma x_k + 2a_2 = 0 \qquad (3.3\text{–}8b)$$
$$\Sigma x_k^3 + a_1 \Sigma x_k^2 + a_2 \Sigma x_k + 3a_3 = 0 \;\ldots \qquad (3.3\text{–}8c)$$

for equation 3.3–1 and

$$\Sigma y_k + A_1 = 0 \qquad (3.3\text{–}9a)$$
$$\Sigma y_k^2 + A_1 \Sigma y_k + 2A_2 = 0 \qquad (3.3\text{–}9b)$$
$$\Sigma y_k^3 + A_1 \Sigma y_k^2 + A_2 \Sigma y_k + 3A_3 = 0 \;\ldots \qquad (3.3\text{–}9c)$$

The Symmetry of Equations

for equation 3.3-3. For example, if the objective is for the $A_1 y^{n-1}$ term in equation 3.3-3 to disappear (e.g., transformation of the general cubic equation 3.3-4 to the special cubic equation 3.3-6), then $\Sigma y_k = 0$ by equation 3.3-9a. This can be achieved by a transformation

$$y_k = \alpha_0 + x_k \qquad (3.3\text{-}10)$$

where $\alpha_1 = 1$. Summing equation 3.3-10 over k gives

$$\Sigma y_k = n\alpha_0 + \Sigma x_k \qquad (3.3\text{-}11)$$

for an equation of degree n. Setting $\Sigma y_k = 0$ and using equation 3.3-8a gives $\alpha_0 = a_1/n$ which gives equation 3.3-7 for the cubic equation where $n = 3$.

In the solution of the general quintic equation a Tschirnhausen transformation is used with the objective of converting the general quintic equation

$$x^5 + a_1 x^4 + a_2 x^3 + a_3 x^2 + a_4 x + a_5 = 0 \qquad (3.3\text{-}12)$$

to the so-called *principal quintic equation*

$$y^5 + A_3 y^2 + A_4 y + A_5 = 0 \qquad (3.3\text{-}13)$$

with the objective of making the y^4 and y^3 terms disappear, i.e., $A_1 = A_2 = 0$. This requires a transformation

$$y_k = \alpha_0 + \alpha_1 x_k + \alpha_2 x_k^2 \qquad (3.3\text{-}14)$$

Squaring equation 3.3-14 gives

$$y_k^2 = \alpha_0^2 + 2\alpha_0 \alpha_1 x_k + \alpha_1^2 x_k^2 + 2\alpha_0 \alpha_2 x_k^2 + 2\alpha_1 \alpha_2 x_k^3 + \alpha_2^2 x_k^4$$

$$(3.3\text{-}15)$$

Summing equations 3.3-14 and 3.3-15 over all of the roots and setting $\Sigma y_k = \Sigma y_k^2 = 0$ to make $A_1 = A_2 = 0$ gives

$$\Sigma y_k = n\alpha_0 + \Sigma x_k \alpha_1 + \Sigma x_k^2 \alpha_2 = 0 \qquad (3.3\text{-}16)$$

$$\Sigma y_k^2 = n\alpha_0^2 + 2\Sigma x_k \alpha_0 \alpha_1 + \Sigma x_k^2 \alpha_1^2 + 2\Sigma x_k^2 \alpha_0 \alpha_2 + 2\Sigma x_k^3 \alpha_1 \alpha_2 + \Sigma x_k^4 \alpha_2^2 = 0 \qquad (3.3\text{-}17)$$

Solving these simultaneous equations leads to a quadratic equation for the ratio $\alpha_1 : \alpha_2$ which thus can be solved by adjoining a square root. Thus a Tschirnhausen transformation to make $A_1 = A_2 = 0$ is therefore possible and requires only adjoining one square root.

Now consider a Tschirnhausen transformation of equation 3.3-1 to 3.3-3 where the y^{n-1}, y^{n-2}, and y^{n-3} terms are made to vanish by setting $A_1 = A_2 = A_3 = 0$. In the case of the general quintic equation (equation 3.3-12), this leads to the Bring-Jerrard equation

$$y^5 - A_4 y + A_5 = 0 \qquad (3.3\text{-}18).$$

Such a Tschirnhausen transformation uses a relationship
$$y_k = \alpha_0 + \alpha_1 x_k + \alpha_2 x_k^2 + \alpha_3 x_k^3 + \alpha_4 x_k^4 \qquad (3.3-19).$$
The requirement $A_1 = 0 \Rightarrow \Sigma y_k = 0$ leads to the condition
$$n\alpha_0 + \alpha_1 \Sigma x_k + \alpha_2 \Sigma x_k^2 + \alpha_3 \Sigma x_k^3 + \alpha_4 \Sigma x_k^4 = 0 \qquad (3.3-20).$$
The additional requirements $A_2 = A_3 = 0 \Rightarrow \Sigma y_k^2 = \Sigma y_k^3 = 0$ lead to two homogeneous equations of degrees 2 and 3, namely

$$\sum_{\lambda=0}^{4} \sum_{\mu=0}^{4} (\Sigma x_k^{\lambda+\mu}) \alpha_\lambda \alpha_\mu = 0 \qquad (3.3-21a)$$

$$\sum_{\lambda=0}^{4} \sum_{\mu=0}^{4} \sum_{\nu=0}^{4} (\Sigma x_k^{\lambda+\mu+\nu}) \alpha_\lambda \alpha_\mu \alpha_\nu = 0 \qquad (3.3-21b)$$

Solving equation 3.3–20 for α_0 in terms of α_m ($m = 1,2,3,4$) and substituting this value in equations 3.3–21 gives the following equations:

$$\sum_{\lambda=1}^{4} \sum_{\mu=1}^{4} B_{\lambda\mu} \alpha_\lambda \alpha_\mu = 0 \qquad (3.3-22a)$$

$$\sum_{\lambda=1}^{4} \sum_{\mu=1}^{4} \sum_{\nu=0}^{4} C_{\lambda\mu\nu} \alpha_\lambda \alpha_\mu \alpha_\nu = 0 \qquad (3.3-22b)$$

It can be shown that equation 3.3–22a can be satisfied by adjoining two square roots. Similarly, equation 3.3–22b can be shown to require a cubic equation to determine to determine some of the α_m values. This Tschirnhausen transformation thus can become very difficult to carry out in practice. Cayley[5] gives this Tschirnhausen transformation in detail for converting the general quintic (3.3–12) to the Bring-Jerrard quintic (3.3–18), but it is very complicated and tedious.

The original method for solution of the general quintic equation using elliptic functions reported by Hermite in 1858 required this very complicated Tschirnhausen transformation for converting the general quintic (3.3–12) to the Bring-Jerrard quintic (3.3–18), which could readily be solved using elliptic functions (Chapter 6). However, an easier procedure for solving the general quintic equation uses the much easier Tschirnhausen transformation of the general quintic (3.3–12) to the principal quintic (3.3–13) in which only A_1 and A_2 are zero. Then a second Tschirnhausen transformation of a different type is

The Symmetry of Equations

used which is derived from the special symmetry of the icosahedron[12] as described in Section 6.3. This Tschirnhausen transformation converts the principal quintic (3.3–13), namely an example of an equation (3.3–1) where $a_1 = a_2 = 0$, to an equation of the following type known as the Brioschi quintic:

$$z^5 - 10Zz^3 + 45Z^2z - Z^2 = 0 \qquad (3.3-23)$$

The Brioschi quintic is not only an example of an equation of the type 3.3–3 where $A_1 = A_3 = 0$ but also has coefficients which can be described by a single parameter conventionally designated as Z. The analogue of equation 3.3–2 relating the y_ks of the principal quintic equation 3.3–13 in y to the z_k's of the Brioschi quintic equation 3.3–23 in z can be derived by solving the following equation for z_k:

$$y_k = \frac{\lambda + \mu z_k}{(z_k^2/Z) - 3} \qquad (3.3-24)$$

Note that solving equation 3.3–24 for z_k in terms of y_k does not give a polynomial for y_k in terms of z_k. Thus the transformation of the principal quintic (3.3–13) to the Brioschi quintic (3.3–23) is a more general type of Tschirnhausen transformation than those defined by equations 3.3–1 to 3.3–3 where the analogue of equation 3.3–2 is *not* a simple polynomial.

---

[12] R. B. King and E. R. Canfield, Icosahedral Symmetry and the Quintic Equation, *Comput. and Math. with Appl.*, **24**, 13–28 (1992).

Chapter 4

Elliptic Functions

4.1 Elliptic Functions by the Generalization of Radicals

The previous chapter discusses the application of group theory to the symmetry inherent in algebraic equations with the idea of relating the Galois group of an equation to the methods required for its solution. Thus a consequence of the solubility of all permutation groups of degrees 2, 3, 4 is the fact that radicals of the general type $\sqrt[n]{a}$ ($n = 2$ and/or 3) are sufficient for solving any quadratic, cubic, or quartic algebraic equation (i.e., equations of degrees 2, 3, and 4, respectively). These radicals $\sqrt[n]{a}$ can be evaluated using logarithms as follows where $\ln a = \log_e a$:

$$\sqrt[n]{a} \equiv a^{1/n} = \ln^{-1}\left(\frac{1}{n} \ln a\right) = \exp\left(\frac{1}{n} \ln a\right) \qquad (4.1\text{--}1)$$

The natural logarithm required for equation 4.1–1 arises from the following indefinite integral:

$$\ln x = \int \frac{dx}{x} = \int \frac{dx}{\sqrt{x^2}} \qquad (4.1\text{--}2)$$

Thus the logarithm is an example of a transcendental function that can be expressed in the general form

$$f(x) = \int \frac{dx}{\sqrt{\mathcal{P}(x)}} \qquad (4.1\text{--}3)$$

in which $\mathcal{P}(x)$ is a polynomial. In the case of the natural logarithm the polynomial $\mathcal{P}(x)$ is the degree 2 polynomial x^2.

In some cases, notably the solution of the general cubic equation with three real roots (Section 5.1), it is more convenient to substitute trigonometric functions for radicals, since the necessary radicals would require the difficult calculation of real numbers by taking cube roots of complex numbers. These

Elliptic Functions

trigonometric functions can be generated by inversion of integrals of the type 4.1-3 where $\mathcal{P}$ is also a degree 2 polynomial. Thus consider the expression

$$\int \frac{x+b}{\sqrt{a^2-x^2}} dx = \int \frac{x}{\sqrt{a^2-x^2}} dx + \int \frac{b}{\sqrt{a^2-x^2}} dx$$

$$= -\sqrt{a^2-x^2} + b \int \frac{dx}{\sqrt{a^2-x^2}} = -\sqrt{a^2-x^2} + bu \qquad (4.1-4)$$

However

$$\int \frac{dx}{\sqrt{a^2-x^2}} = u = \sin^{-1} x/a \qquad (4.1-5).$$

Inversion of equation 4.1-5 gives $x = a \sin u$. Setting $a = 1$ gives

$$\sin^{-1} x = \int \frac{dx}{\sqrt{1-x^2}} \qquad (4.1-6)$$

The above analysis shows how logarithms and trigonometric functions, which are sufficient to solve all quadratic, cubic, and quartic equations, can be derived from integrals of the type 4.1-3 in which $\mathcal{P}(x)$ is a second degree polynomial. However, radicals, and thus logarithms and trigonometric functions, are no longer sufficient for solution of the general quintic equation, i.e., a general equation of degree 5, since the corresponding Galois group of the general quintic equation is the alternating group A_5, which is a simple group. In order to solve the general quintic equation the concept of radicals must be generalized to what have been called *ultraradicals*, i.e., $\sqrt[*]{a}$ for the real zero of $x^5 + x - a$.[1] Such ultraradicals for the general quintic equation turn out to be elliptic functions, which can be derived from integrals[2] of the type 4.1-3 in which the polynomial $\mathcal{P}(x)$ is of degree 3 or 4 analogous to the derivation of logarithms and trigonometric functions from such integrals (4.1-3) in which the polynomial $\mathcal{P}(x)$ is of degree 2. Thus consider the following expression which

---

[1] I. Stewart, *Galois Theory*, Chapman and Hall, London, 1973.
[2] H. E. Rauch and A. Lebowitz, *Elliptic Functions, Theta Functions, and Riemann Surfaces*, Williams and Wilkins, Baltimore, MD, 1973.

is the cubic analogue to equation 4.1–4:

$$\int \frac{x^2 + b}{\sqrt{x^3+ax+1}}dx =$$
$$\int \frac{(x^2 + a/3)}{\sqrt{x^3+ax+1}}dx + \int \frac{(b - a/3)}{\sqrt{x^3+ax+1}}dx$$
$$= \frac{2}{3}\sqrt{x^3+ax+1} + \int \frac{(b - a/3)}{\sqrt{x^3+ax+1}}dx \qquad (4.1-7)$$

Equation 4.1–7 for $b - a/3 \neq 0$ gives an integral whose evaluation for $a^3 \neq -27/4$ in terms of familiar elementary functions (algebraic, logarithmic, exponential, and trigonometric) was unsuccessfully sought for more than a century after Newton's and Leibniz's basic discoveries of calculus By the end of the 18th century the leading mathematicians such as Euler, Lagrange, and Legendre were rather sure that integrals of the type 4.1–3 where $\mathcal{P}(x)$ is a degree 3 polynomial represented a new type of transcendental function. Such integrals, even before they were understood, were called *elliptic integrals* since related integrals arise in the determination of the arc length of an ellipse.

The original work on elliptic functions was done in the late 1820s by Abel and Jacobi using an integral of the type 4.1–3 where $\mathcal{P}(x)$ is the *fourth* degree polynomial

$$\mathcal{P}(x) = (1-x^2)(1-k^2x^2) \qquad (4.1-8).$$

This fourth degree polynomial for $\mathcal{P}(x)$ was chosen because of its resemblance to the second degree polynomial $\mathcal{P}(x) = 1 - x^2$ found in the integral expression for $\sin^{-1}x$ (equation 4.1–6). This also relates to extension of the single periodicity of the trigonometric functions to the double periodicity of the elliptic functions with the *modulus k* being the ratio between the two periods of the elliptic functions (Section 4.2).

The first elliptic functions to be defined used the following definite integral to define a new function $f(u)$:

$$u = \int_{x=0}^{x=f(u)} \frac{dx}{\sqrt{(1-x^2)(1-k^2x^2)}} \qquad (4.1-9)$$

Elliptic Functions

Jacobi made the trigonometric substitution $x = \sin \phi$ so that

$$u = \int_0^\phi \frac{\cos\theta\, d\theta}{\sqrt{\cos^2\theta(1-k^2\sin^2\theta)}} = \int_0^\phi \frac{d\theta}{\sqrt{1-k^2\sin^2\theta}} \qquad (4.1\text{–}10)$$

The variable ϕ was then called the amplitude of u so that $x = f(u) = \sin \phi = \sin$ amplitude u abbreviated first by Jacobi to sin am u and subsequently by Gudermann to sn u, which is the most commonly used notation today. The other Jacobi elliptic functions are defined as follows:

$$\text{cn } u = \cos\phi = \sqrt{1 - \text{sn}^2 u} \qquad (4.1\text{–}11a)$$
$$\text{dn } u = \sqrt{1 - k^2 \text{sn}^2 u} \qquad (4.1\text{–}11b)$$

The explicit dependence of sn u on the modulus k can be introduced by writing $\text{sn}(u,k)$. The *complementary modulus* k' is defined by $k' = \sqrt{1 - k^2}$. Note that sn $(0,k) = 1$ for all k.

The Jacobi elliptic functions sn u, cn u, and dn u defined by equations 4.1–10 and 4.1–11 use an integral of the type 4.1–3 where $\mathcal{P}(x)$ is a quartic polynomial. It is now instructive to relate this type of integral to similar integrals where $\mathcal{P}(x)$ is only a cubic polynomial. Consider the use of an elliptic integral to define an elliptic curve by first defining a general quartic polynomial by the equation

$$w^2 = \mathcal{P}(z) = a_4 z^4 + a_3 z^3 + a_2 z^2 + a_1 z + a_0$$
$$= a_4(z - z_1)(z - z_2)(z - z_3)(z - z_4) \qquad (4.1\text{–}12)$$

in which $a_0, \ldots, a_4$ are arbitrary complex constants and $z_1, \ldots, z_4$ are the roots (assumed distinct) of the polynomial in equation 4.1–12. If z is a complex variable, then equation 4.1–12 defines w as a two-valued algebraic function of z. Abel then defined an *elliptic integral* to be an integral of the form $\int R(z,w)dz$ in which $R(z,w)$ is a rational function of the complex variables z and w. The locus of all values of z and w satisfying equation 4.1–12 can appear as a two-dimensional locus in the four-dimensional space of the two complex variables z

and w. Equation 4.1–12 may be regarded as defining an algebraic curve so that a function of the form $R(z,w)$ is then a function on this curve. Because integrals of the type 4.1–3 where $\mathcal{P}(x)$ is a cubic or quartic polynomial are called elliptic integrals, this curve is called an *elliptic curve*.

The most general elliptic curve defined by equation 4.1–12 would appear to depend on four parameters, namely z_k ($1 \le k \le 4$). However, for the purposes of integration only one complex parameter is required as can be seen by putting equation 4.1–12 into a standard form, called the normal form. In order to put equation 4.1–12 into the normal form, first define

$$z' = \frac{(z - z_4)(z_3 - z_2)}{(z - z_2)(z_3 - z_4)} \qquad (4.1\text{–}13)$$

Applying equation 4.1–13 to $z' = z_1$ gives

$$\frac{1}{\lambda} = \frac{(z_1 - z_4)(z_3 - z_2)}{(z_1 - z_2)(z_3 - z_4)} \qquad (4.1\text{–}14)$$

where λ is called the *cross-ratio* of $z_1,\ldots,z_4$. Similar applications of equation 4.1–13 take z_2 to ∞, z_3 to 1, and z_4 to 0. Now solve for z as a function of z' and substitute z' for z in the right-hand side of equation 4.1–12 to give

$$w^2\left[1 - z'\left(\frac{z_3 - z_4}{z_3 - z_2}\right)\right]^4 = a_4' z'(1 - z')(1 - \lambda z') \qquad (4.1\text{–}15)$$

where $a_4' = (z_2 - z_4)^2 (z_3 - z_4)^2 \left(\frac{z_1 - z_4}{z_3 - z_2}\right) a_4$ \qquad (4.1–16)

Now define

$$w' = w\left[1 - z'\left(\frac{z_3 - z_4}{z_3 - z_2}\right)\right]^2 \quad \text{and} \quad w'' = \frac{2w'}{\sqrt{a_4'}} \qquad (4.1\text{–}17)$$

Then
$$w'^2 = a_4' z'(1 - z')(1 - \lambda z') \qquad (4.1\text{–}18a)$$
and $\quad w''^2 = 4z'(1 - z')(1 - \lambda z') \qquad (4.1\text{–}18b)$

Dropping all primes gives *Riemann's normal form* of the elliptic curve. Thus if

Elliptic Functions

an elliptic curve is defined by (4.1–12), its normal form can be written by computing the cross-ratio of the four roots of the polynomial. If the arbitrary order in which the roots are taken is changed, λ will be replaced by $\frac{1}{\lambda}$, $1 - \lambda$, $\frac{1}{1 - \lambda}$, $\frac{\lambda - 1}{\lambda}$, or $\frac{\lambda}{\lambda - 1}$. Note also that in Riemann's normal form (4.1–18) an elliptic curve in (4.1–12) is defined by a cubic equation rather than the original quartic equation. Furthermore, suppose that a_4 in (4.1–12a) is zero so that the resulting quartic equation becomes a cubic equation with three distinct roots. This still defines an elliptic curve, and the transformation to Riemann's normal form is nearly trivial since there is no need to send one root to ∞ by use of the transformation in (4.1–13). Thus if the roots of the cubic equation are z_1, z_2, and z_3, then the corresponding linear transformation $z' = \frac{z - z_1}{z_2 - z_1}$ leaves ∞ fixed and has the following additional effects:

$$z_1 \to 0, \ z_2 \to 1, \ z_3 = \frac{1}{\lambda} = \frac{z_3 - z_1}{z_2 - z_1} \tag{4.1–19}$$

In this way the cubic can be viewed as a degenerate quartic with one zero at ∞. The two branches of w arising from the square root in solving equation 4.1–12 for w coalesce at three zeros and one infinity in the cubic as a degenerate quartic rather than at four zeros in the case of a true quartic. *This shows that cubic and quartic polynomials as $\mathcal{P}(x)$ in equation 4.1–3 give the same kind of transcendental functions, namely the elliptic functions.* Any other degenerate form of the quartic equation on the right side of equation 4.1–12, such as a double root or reduction to a quadratic or linear polynomial, leads back to more elementary functions, namely algebraic, logarithmic, exponential, and/or trigonometric functions.

The differential equations derived from Riemann's normal forms (equations 4.1–18) lead to another integral of the type 4.1–3 analogous to equation 4.1–9 but where $\mathcal{P}(x)$ is now a cubic polynomial, namely

$$u = \int_{x=0}^{x=f(u)} \frac{dx}{\sqrt{4x^3 - g_2 x - g_3}} \tag{4.1–20}$$

The function $f(u)$ defined by inversion of this integral is called the *Weierstrass ℘-function* and is conventionally designated as $\wp(u)$ or more precisely $\wp(u;g_2,g_3)$ to emphasize the role played by the parameters g_2 and g_3. The function $\wp(u)$ turns out to be single-valued and analytic except for an infinity of poles of second order. Since the inversion of the general integral 4.1–20 is a formidable problem, an alternative approach to the Weierstrass elliptic functions based on a study of doubly periodic functions discussed below is more useful.

4.2 Elliptic Functions as Doubly Periodic Functions

Let $f(z)$ be a single-valued function which is analytic on the complex plane $z = x + iy$ except for isolated singularities such as *poles*, which are defined as points where $f(z) = \infty$. A *period, p,* of $f(z)$ is a complex number such that

$$f(z) = f(z + p) \qquad (4.2\text{--}1)$$

for all z for which $f(z)$ is analytic. A function which has one (non-zero) period has an infinity of periods, e.g., np for all integers n. Let Ω be the set of all points in the Argand (complex) plane, $z = x + iy$, corresponding to periods of a function $f(z)$. Periodic functions can be of the following types:

(1) The trivial case where $f(z)$ is a constant so that Ω is the whole complex plane;

(2) The case where $f(z)$ is a single periodic function, such as the trigonometric functions $\sin z$, $\cos z$, $\tan z$, etc., so that Ω is a system of equidistant points on a straight line through the origin of the complex plane;

(3) The case where $f(z)$ is a doubly periodic function so that Ω is a point lattice formed by intersections of two families of equidistant parallel lines.

The point lattice of a doubly periodic function $f(z)$ in the Argand (complex) plane may also be generated by the repetition of congruent parallelograms. Consider one such parallelogram with one of its vertices at 0 and let the other three vertices be 2ω, $2\omega'$, and $2\omega + 2\omega'$. Then 2ω and $2\omega'$ are a pair of *primitive periods* of $f(z)$, and all periods of $f(z)$ have the following form where m and n are integers:

$$2\omega_{m,n} = 2m\omega + 2n\omega' \qquad (4.2\text{--}2)$$

Clearly the ratio ω'/ω is not real. The primitive periods are conventionally

Elliptic Functions

chosen such that $\text{Im}(\omega'/\omega) > 0$ where $\text{Im } z$ refers to the imaginary part of the complex number z.

Two points of the complex plane corresponding to the variable z of a doubly periodic function are said to be *congruent* if they differ by a period. A connected set of points is called a *fundamental region* if every point of the plane is congruent to exactly one point of the set. The fundamental region is chosen to be a parallelogram with two sides and the vertex at which they intersect being considered as part of the parallelogram and the other two sides and three vertices not forming part of the parallelograms. If z_0 is fixed the points

$$z = z_0 + 2\xi\omega + 2\eta\omega' \qquad (4.2\text{--}3)$$

where $0 \le \xi, \eta < 1$ form the *fundamental period parallelogram*. Any parallelogram obtained from the fundamental period parallelogram by translation by an integral number of periods is a *period parallelogram*, or a *mesh*, which corresponds to a set of points

$$z = z_0 + 2(m + \xi)\omega + 2(n + \eta)\omega' \qquad (4.2\text{--}4)$$

with a fixed pair of integers $\{m,n\}$. Figure 4–1 shows two different ways of choosing period parallelograms for a given period lattice Ω. Since a doubly-periodic function assumes the same value at congruent points, it is sufficient to describe the behavior of such a function in any single mesh. Furthermore, since $f(z)$ has only isolated singularities and isolated zeros, the fundamental period parallelogram (i.e., z_0 in equation 4.2–3) can be chosen so that no singularities or zeros of $f(z)$ lie on the boundary of such a mesh. A mesh with these properties can be called a *cell*.

An *elliptic function* is defined as a single-valued doubly-periodic analytic function whose only possible singularities in the finite part of the plane are poles. Such elliptic functions have the following properties:[3]

(1) Every nonconstant elliptic function has poles. Otherwise, if $f(z)$ has no poles in a mesh, it is bounded there and hence in the entire complex plane so that it must be a constant by Liouville's theorem.

(2) An elliptic function has only finite numbers of poles and zeros in any mesh, if it does not vanish identically. Infinities of poles or zeros in a mesh imply the

---

[3] A. W. Erdélyi, W. Magnus, F. Oberhettinger, and F. C. Tricomi, *Higher Transcendental Functions*, McGraw-Hill, New York, 1953, Chapter 13, part 2.

presence of singularities other than poles. The number of poles in a cell, each pole counted according to its multiplicity, is called the *order* of an elliptic function. The set of poles and zeros in a given cell is called an *irreducible set*.

(3) The sum of the residues of an elliptic function at its poles in any cell is zero. If C is the boundary of the cell, the sum of the residues $\int_C f(z)dz$ is zero since the integrals along opposite sides cancel.

(4) There is no elliptic function of order one. Such a function would have exactly one simple pole in each cell which is inconsistent with a residue sum of zero (property 3 above).

(5) An elliptic function of order r assumes every value exactly r times in any mesh counting multiplicity.

(6) The sum of an irreducible set of zeros is congruent to the sum of an irreducible set of poles with each zero and pole being repeated according to its multiplicity.

(7) Two elliptic functions, which have the same periods, the same poles, and the same principal parts at each pole differ by a constant.

(8) The quotient of two elliptic functions whose periods, poles, and zeros are the same (including the multiplicities of the poles and zeros) is a constant.

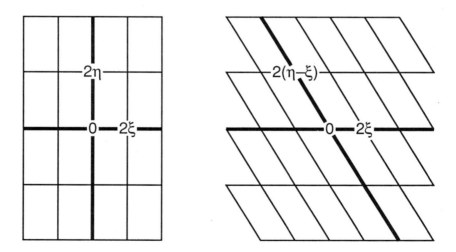

Figure 4–1: Two different ways of choosing period parallelograms for a given point-lattice Ω in the complex (Argand) plane.

Elliptic Functions

All elliptic functions with the same periods 2ω and $2\omega'$ form a field K (Section 3.1). Thus any rational function (with constant coefficients) of elliptic functions with the same periods belongs to K. Moreover, the derivative of any function in K belongs also to K so that K is a *differential field*. However, an integral of a function of K does not necessarily belong to K. Although $\{2\omega,2\omega'\}$ is a pair of *primitive* periods for *some* functions in K and a pair of periods for all functions in K, it is not necessarily a pair of primitive periods for all functions in K. Any two elliptic functions f and g of the field K can be shown to be connected by an algebraic equation $\mathcal{P}(f,g) = 0$ where $\mathcal{P}(x,y)$ is a polynomial with constant coefficients and the algebraic curve $\mathcal{P}(x,y)$ is unicursal. In addition, any elliptic function f satisfies an algebraic differential equation of the first order $\mathcal{P}(f,f') = 0$ where $\mathcal{P}(x,y)$ again is a polynomial with constant coefficients and genus zero. Any elliptic function, $f(z)$, can be shown to satisfy an algebraic addition theorem $\mathcal{A}[f(u),f(v),f(u+v)] \equiv 0$ where $\mathcal{A}(x,y,z)$ is a polynomial whose coefficients are independent of u and v. Conversely it can be shown that a single-valued analytic function of z which satisfies an algebraic addition theorem of the above form is either a rational function of z, a rational function of $e^{\lambda z}$ for some λ, or else an elliptic function.

The simplest nontrivial elliptic functions are functions of order two. These may be of either of the following two types:

(1)   One double pole with residue zero in each cell leading to the Weierstrass type of elliptic functions discussed below;

(2)   Two simple poles with residues equal in magnitude but opposite in sign in each cell leading to the Jacobi elliptic functions discussed above.

These considerations lead to the definition of the Weierstrass elliptic functions in terms of their periodicity rather than by inversion of the integral in (4.1–20). The Weierstrass function $\wp(z) = \wp(z|\omega,\omega')$ is thus uniquely defined as an elliptic function of order two with periods 2ω, $2\omega'$ and a double pole at $z = 0$ with principal part $1/z^2$ around which $\wp(z) - 1/z^2$ is analytic and vanishes at $z = 0$. An analytical expression for $\wp(z)$ can be obtained by constructing a meromorphic function with double poles and principal parts $\dfrac{1}{(z-w)^2}$ at all points $w = w_{mn}$. The partial fraction expansion of such a function can be obtained by the following equation:

$$f(z) = \frac{1}{z^2} + \sum_{(m,n)\neq(0,0)} \left[\frac{1}{(z-w)^2} - \frac{1}{w^2} \right] \qquad (4.2-5)$$

Moreover $f(z) - \frac{1}{z^2}$ vanishes at $z = 0$. By rearranging the series it can be proved that $f(z + 2\omega) = f(z) = f(z + 2\omega')$ leading to the conclusion that $f(z) = \wp(z)$ or

$$\wp(z) = \wp(z \mid \omega,\omega') = \frac{1}{z^2} + \sum_{(m,n)\neq(0,0)} \left[\frac{1}{(z-2m\omega-2n\omega')^2} - \frac{1}{(2m\omega+2n\omega')^2} \right]$$

$$(4.2-6)$$

The function $\wp(z)$ is an even function of z.

The derivatives and integrals of the Weierstrass function $\wp(z)$ are of interest.[3] Differentiating $\wp(z)$ gives the following equation:

$$\wp'(z) = -\frac{2}{z^3} - 2 \sum_{(m,n)\neq(0,0)} \frac{1}{(z-w)^3} = -2 \sum_{\text{all }(m,n)} \frac{1}{(z-w)^3} \qquad (4.2-7)$$

Integrating $\wp(z)$ term by term gives the *Weierstrass zeta function*, which is a meromorphic function with simple poles, i.e.,

$$\zeta(z) = \zeta(z \mid \omega,\omega') = \frac{1}{z} + \sum_{(m,n)\neq(0,0)} \left[\frac{1}{z-w} + \frac{1}{w} + \frac{z}{w^2} \right] \qquad (4.2-8a)$$

$$\wp(z) = -\zeta'(z) \qquad (4.2-8b)$$

Note that the function $\zeta(z)$ is an odd function of z (i.e., $\zeta(z) = -\zeta(-z)$) and is *not* doubly-periodic and therefore *not* an elliptic function. The variables η and η' are used to express the lack of periodicity of $\zeta(z)$ as follows:

$$\zeta(z + 2\omega) = \zeta(z) + 2\eta \qquad (4.2-9a)$$
$$\zeta(z + 2\omega') = \zeta(z) + 2\eta' \qquad (4.2-9b)$$

Integrating $\zeta(z)$ around a cell gives *Legendre's relation*, namely

$$\eta\omega' - \eta'\omega = 1/2\ i\pi \qquad (4.2-10)$$

The *Weierstrass sigma function* is an entire function whose logarithmic

Elliptic Functions

derivative is the zeta function, i.e.,

$$\sigma(z) = \sigma(z|\omega,\omega') = z \prod_{(m,n)\neq(0,0)} \left\{ \left(1 - \frac{z}{w}\right) \exp\left[\frac{z}{w} + \frac{1}{2}\left(\frac{z}{w}\right)^2\right] \right\} \quad (4.2\text{–}11a)$$

$$\zeta(z) = \frac{\sigma'(z)}{\sigma(z)} \quad (4.2\text{–}11b)$$

$$\wp(z) = \frac{-d^2[\ln \sigma(z)]}{dz^2} = \frac{\sigma'^2(z) - \sigma(z)\sigma''(z)}{\sigma^2(z)} \quad (4.2\text{–}11c)$$

Using the abbreviations

$$g_2 = 60 \sum_{(m,n)\neq(0,0)} \frac{1}{w^4} \quad \text{and} \quad g_3 = 140 \sum_{(m,n)\neq(0,0)} \frac{1}{w^6} \quad (4.2\text{–}12)$$

the power series expansion of $\sigma(z)$ and the Laurent series expansions of $\zeta(z)$, $\wp(z)$ and $\wp(z)$ in the neighborhood of the origin are as follows:

$$\sigma(z) = z - \frac{g_2 z^5}{2^4 \cdot 3 \cdot 5} - \frac{g_3 z^7}{2^3 \cdot 3 \cdot 5 \cdot 7} - \frac{g_2^2 z^9}{2^9 \cdot 3^2 \cdot 5 \cdot 7} - \frac{g_2 g_3 z^{11}}{2^7 \cdot 3^2 \cdot 5^2 \cdot 7 \cdot 11} - \dots \quad (4.2\text{–}13a)$$

$$\zeta(z) = \frac{1}{z} + \frac{g_2 z^3}{2^2 \cdot 3 \cdot 5} - \frac{g_3 z^5}{2^2 \cdot 5 \cdot 7} - \frac{g_2^2 z^7}{2^4 \cdot 3 \cdot 5^2 \cdot 7} + \dots \quad (4.2\text{–}13b)$$

$$\wp(z) = \frac{1}{z^2} + \frac{g_2 z^2}{2^2 \cdot 5} + \frac{g_3 z^4}{2^2 \cdot 7} + \frac{g_2^2 z^6}{2^4 \cdot 3 \cdot 5^2} + \dots \quad (4.2\text{–}13c)$$

$$\wp'(z) = -\frac{2}{z^3} + \frac{g_2 z}{2 \cdot 5} + \frac{g_3 z^3}{7} + \frac{g_2^2 z^5}{2^3 \cdot 5^2} + \dots \quad (4.2\text{–}13d)$$

Formulas with the Weierstrass elliptic functions can be expressed more symmetrically using the notation $\omega_1 = \omega$, $\omega_2 = -\omega - \omega'$, and $\omega_3 = \omega'$ and $\eta_a = \zeta(\omega_a)$ ($a = 1,2,3$) leading to the following equations ($a = 1,2,3$):

$$\zeta(z + 2\omega_a) = \zeta(z) + 2\eta_a \quad (4.2\text{–}14a)$$

$$\sigma(z + 2\omega_a) = -\sigma(z)\exp[2\eta_a(z+\omega_a)] \quad (4.2\text{–}14b)$$

The following three functions can then be defined for $a = 1,2,3$:

$$\sigma_a(z) = \frac{\sigma(z + \omega_a)}{\sigma(\omega_a)} \exp(-z\eta_a) \qquad (4.2\text{--}15)$$

The function $\wp'(z)$ is an odd elliptic function of order three with periods $2\omega_a$ ($a = 1,2,3$) with three zeros in every cell. Now $\wp'(-\omega_a) = \wp'(\omega_a)$ since $\wp'$ has a period $2\omega_a$ and $\wp'(-\omega_a) = -\wp'(\omega_a)$ since $\wp'$ is an odd function of z. Therefore $z = \omega_a$ ($a = 1,2,3$) is seen to be an irreducible set of zeros for $\wp'(z)$ which are customarily designated as

$$e_a = \wp(\omega_a) \quad \text{for } a = 1,2,3 \qquad (4.2\text{--}16)$$

The function $\wp(z) - \wp(\omega_a)$ is an elliptic function of order two with double poles at points congruent to 0 and double zeros at point congruent to ω_a. Since it is of order two, these are the only poles and zeros so that the function $\sqrt{\wp(z) - e_a}$ can be defined as a single-valued function but not necessarily with periods 2ω and $2\omega'$.

The functions $\wp'^2(z)$ and $[\wp(z) - e_1][\wp(z) - e_2][\wp(z) - e_3]$ are both elliptic functions of order six with the same periods (i.e., $2\omega_\alpha$ for $\alpha = 1, 2, 3$), the same irreducible set of double zeros at ω_α ($\alpha = 1, 2, 3$), and a pole of order six at 0 so that their quotient is constant. Using equations 4.2–6 and 4.2–7 an algebraic differential equation for the Weierstrass $\wp$ function can be obtained, i.e.,

$$\wp'^2(z) = 4[\wp(z) - e_1][\wp(z) - e_2][\wp(z) - e_3] \qquad (4.2\text{--}17)$$

An alternative form of this differential equation can be obtained from the observation that $\wp'^2(z) - 4\wp^3(z) + g_2\wp(z)$ is an elliptic function of no more than order six and that all possible poles of this function are congruent to 0. Equations 4.2–13c and 4.2–13d indicate that this function is regular at $z = 0$ and therefore a constant. Furthermore, equations 4.2–13c and 4.2–13d indicate the value of this constant to be $-g_3$ leading to the following alternative differential equation for $\wp'^2$:

$$\wp'^2(z) = 4\wp^3(z) - g_2\wp(z) - g_3 \qquad (4.2\text{--}18)$$

A comparison of the right sides of equations 4.2–17 and 4.2–18 shows that e_1, e_2, and e_3 are the three roots of the algebraic equation $4z^3 - g_2 z - g_3 = 0$ leading to the following relationships between these roots using formulas for the sums of the roots of algebraic equations (compare equations 3.3–8 and 3.3–9):

Elliptic Functions

$$e_1 + e_2 + e_3 = 0 \tag{4.2-19a}$$
$$-4(e_2e_3 + e_3e_1 + e_1e_2) = 0 \tag{4.2-19b}$$
$$4e_1e_2e_3 = g_3 \tag{4.2-19c}$$
$$e_1^2 + e_2^2 + e_3^2 = {}^1\!/_2\, g_2 \tag{4.2-19d}$$
$$e_1^3 + e_2^3 + e_3^3 = {}^3\!/_4\, g_3 \tag{4.2-19e}$$
$$e_1^4 + e_2^4 + e_3^4 = {}^1\!/_8\, g_2^2 \tag{4.2-19f}$$
$$16\,(e_2 - e_3)^2(e_3 - e_1)^2(e_1 - e_2)^2 = g_2^3 - 27 g_3^2 = \Delta \tag{4.2-19g}$$

The quantity Δ is the *discriminant* of the cubic equation in $\wp(z)$ on the right side of (4.2–18).

The Weierstrass elliptic functions can be related to the Jacobi elliptic functions discussed in Section 4.1.[3] Thus consider the Jacobi elliptic function

$$w = \operatorname{sn} u = \operatorname{sn}(u,k) \tag{4.2-20}$$

defined as in equation 4.1–9. The variables u and k of this Jacobi elliptic function bear the following relationships to the variable z and the parameters e_k ($k = 1,2,3$) discussed above for the Weierstrass elliptic functions:

$$e_1 : e_2 : e_3 = (2-k^2) : (2k^2-1) : -(1+k^2) \tag{4.2-21a}$$

$$z = \frac{u}{\sqrt{e_1 - e_3}} \;\Rightarrow\; u = z\sqrt{e_1 - e_3} \tag{4.2-21b}$$

$$k^2 = \frac{e_2 - e_3}{e_1 - e_3} \tag{4.2-21c}$$

$$k'^2 = 1 - k^2 = \frac{e_1 - e_2}{e_1 - e_3} \tag{4.2-21d}$$

The Jacobi elliptic functions can then be expressed as follows using the Weierstrass functions:

$$\operatorname{sn}(u,k) = \frac{\sqrt{e_1 - e_3}}{\sqrt{\wp(z) - e_3}} = \sqrt{e_1 - e_3}\,\frac{\sigma(z)}{\sigma_3(z)} \tag{4.2-22a}$$

$$\operatorname{cn}(u,k) = \frac{\sqrt{\wp(z) - e_1}}{\sqrt{\wp(z) - e_3}} = \frac{\sigma_1(z)}{\sigma_3(z)} \tag{4.2-22b}$$

$$\text{dn}(u,k) = \frac{\sqrt{\wp(z) - e_2}}{\sqrt{\wp(z) - e_3}} = \frac{\sigma_2(z)}{\sigma_3(z)} \qquad (4.2\text{--}22\text{c})$$

The Weierstrass elliptic functions exhibit a number of *addition theorems* of interest in the context of algebraic equations.[3] The addition theorem of the $\wp$-function may be written in several forms, i.e.,

$$\wp(u+v) = \frac{1}{4}\left[\frac{\wp'(u) - \wp'(v)}{\wp(u) - \wp(v)}\right]^2 - \wp(u) - \wp(v) \qquad (4.2\text{--}23\text{a})$$

$$\begin{vmatrix} 1 & \wp(u) & \wp'(u) \\ 1 & \wp(v) & \wp'(v) \\ 1 & \wp(u+v) & -\wp'(u+v) \end{vmatrix} = 0 \qquad (4.2\text{--}23\text{b})$$

$$\wp(u+v) = \wp(u) - \frac{1}{2}\frac{\partial}{\partial u}\left[\frac{\wp'(u) - \wp'(v)}{\wp(u) - \wp(v)}\right]$$
$$= \wp(v) - \frac{1}{2}\frac{\partial}{\partial v}\left[\frac{\wp'(u) - \wp'(v)}{\wp(u) - \wp(v)}\right] \qquad (4.2\text{--}23\text{c})$$

$$\wp(u+v) + \wp(u-v) = 2\wp(u) - \frac{\partial^2}{\partial u^2}\{\ln[\wp(u) - \wp(v)]\} \qquad (4.2\text{--}23\text{d})$$

These addition theorems may be proved by observing that the functions on the two sides of the equation are elliptic functions with the same periods, poles, and principal parts, and have the same value at some specified point. From these addition formulas for $\wp(u)$ the following related formulas can be derived:

$$\wp(2z) = -2\wp(z) + \left[\frac{\wp''(z)}{2\wp'(z)}\right]^2 \qquad (4.2\text{--}24\text{a})$$

$$\wp(\tfrac{1}{2}z) = \wp(z) + \sqrt{\wp(z) - e_2}\sqrt{\wp(z) - e_3} +$$
$$\sqrt{\wp(z) - e_3}\sqrt{\wp(z) - e_1} +$$
$$\sqrt{\wp(z) - e_1}\sqrt{\wp(z) - e_2} \qquad (4.2\text{--}24\text{b})$$

Elliptic Functions

In equation 4.2–24b the square roots are taken so that $z = 0$ is a simple pole with residue unity thereby implying

$$\wp'(z) = -2 \sqrt{\wp(z) - e_1} \sqrt{\wp(z) - e_2} \sqrt{\wp(z) - e_3} \qquad (4.2\text{–}25)$$

Some related expressions can be derived for the Weierstrass zeta and sigma functions, e.g.,

$$\zeta(u+v) = \zeta(u) + \zeta(v) + \frac{1}{2} \frac{\wp'(u) - \wp'(v)}{\wp(u) - \wp(v)} \qquad (4.2\text{–}26a)$$

$$\sigma(u+v)\sigma(u-v) = -\sigma^2(u)\sigma^2(v)[\wp(u) - \wp(v)] \qquad (4.2\text{–}26b)$$

Equations 4.2–26 are not true addition theorems since the right hand side contains functions other than the function on the left hand side. The following equations can be derived from equations 4.2–26:

$$\zeta(z + 2m\omega + 2n\omega') = \zeta(z) + 2m\eta + 2n\eta' \qquad (4.2\text{–}27a)$$

$$\sigma(z + 2m\omega + 2n\omega') = (-1)^{m+n+mn} \sigma(z) \exp[(z+m\omega+n\omega')(2m\eta+2n\eta') \qquad (4.2\text{–}27b)$$

In these equations m and n are integers and η and η' are defined as in (4.2–9). These equations are derived by expressing $\dfrac{\wp'(u) - \wp'(v)}{\wp(u) - \wp(v)}$ in terms of zeta functions and $\wp(u) - \wp(v)$ in terms of sigma functions.

Another function of interest is the quotient $\sigma(nu)/\sigma^{n^2}(u)$, since the case where $n = 5$ (i.e., $\sigma(5u)/\sigma^{25}(u)$) is used in the solution of the quintic equation. A useful formula for this function attributed to Kiepert is given by Schwarz in a set of notes originating from Weierstrass and published in 1893.[4] This set of old notes is a useful source for other relatively obscure information on elliptic functions, particularly those of the Weierstrass type. The general formula for

---

[4]H. A. Schwarz and K. Weierstrass, *Formeln und Lehrsätze zum Gebrauche der Elliptischen Functionen*, Springer, Berlin, 1893, p. 19.

$\sigma(nu)/\sigma^{n^2}(u)$ can be expressed as follows where $\wp^{(n)}$ is the nth derivative of $\wp$:

$$\frac{\sigma(nu)}{\sigma^{n^2}(u)} = \frac{(-1)^{n-1}}{(1!2!3!\ldots(n-1)!)^2} \begin{vmatrix} \wp'(u) & \wp''(u) & \ldots & \wp^{(n-1)}(u) \\ \wp''(u) & \wp'''(u) & \ldots & \wp^{(n)}(u) \\ \cdot & \cdot & \ldots & \cdot \\ \wp^{(n-1)}(u) & \wp^{(n)}(u) & \ldots & \wp^{(2n-3)}(u) \end{vmatrix}$$

(4.2–28)

The proof of this rather bizarre formula is beyond the scope of this book but can be found in the notes compiled by Schwarz.[4]

4.3 Theta Functions

Theta functions[5,6] are periodic functions that can be represented by series whose convergence is extraordinarily rapid. They are important in providing the best means for the numerical computation of elliptic functions. For this reason theta functions are useful in the solution of algebraic equations requiring the use of elliptic functions.

Consider a Weierstrass function $\wp(z|\omega,\omega')$ and define $\tau = \omega'/\omega$ selecting ω and ω' so that Im $\tau > 0$. Furthermore define $q = e^{i\pi\tau} = e^{i\pi\omega'/\omega}$ and $v = z/2\omega$. Then the four theta functions can be defined as follows using the designations of Erdelyi et al[3]:

$$\theta_1(v) = \theta_1(v,q) = \theta_1(v|\tau) = 2\sqrt[4]{q} \sum_{n=0}^{\infty} (-1)^n q^{n(n+1)} \sin[(2n+1)\pi v] \quad (4.3\text{–}1a)$$

$$\theta_2(v) = \theta_2(v,q) = \theta_2(v|\tau) = 2\sqrt[4]{q} \sum_{n=0}^{\infty} q^{n(n+1)} \cos[(2n+1)\pi v] \quad (4.3\text{–}1b)$$

---

[5] K. Chandrasekharan, *Elliptic Functions*, Springer, Berlin/Heidelberg, 1985, Chapter 5.
[6] D. Mumford, *Tata Lectures on Theta I*, Birkhäuser, Boston, 1983.

Elliptic Functions

$$\theta_3(v) = \theta_3(v,q) = \theta_3(v|\tau) = 1 + 2 \sum_{n=0}^{\infty} q^{n^2}\cos(2n\pi v) \qquad (4.3\text{--}1c)$$

$$\theta_4(v) = \theta_4(v,q) = \theta_4(v|\tau) = 1 + 2 \sum_{n=0}^{\infty} (-1)^n q^{n^2}\cos(2n\pi v) \qquad (4.3\text{--}1d)$$

These series converge for all (complex) v and all q defined as above. The factor q^{n^2} leads to extremely rapid convergence. The function θ_1 is an odd function ($\theta_1(-v) = -\theta_1(v)$) whereas the functions θ_2, θ_3, and θ_4 are even functions ($\theta_n(v) = \theta_n(-v)$ for $n = 2, 3, 4$). The trigonometric functions in the theta series can be converted into exponential functions leading to the following[3]:

$$\theta_1(v) = i \sum_{n=-\infty}^{\infty} (-1)^n q^{(n-1/2)^2} e^{(2n-1)i\pi v} \qquad (4.3\text{--}2a)$$

$$\theta_2(v) = \sum_{n=-\infty}^{\infty} q^{(n-1/2)^2} e^{(2n-1)i\pi v} \qquad (4.3\text{--}2b)$$

$$\theta_3(v) = \sum_{n=-\infty}^{\infty} q^{n^2} e^{2ni\pi v} \qquad (4.3\text{--}2c)$$

$$\theta_4(v) = \sum_{n=-\infty}^{\infty} (-1)^n q^{n^2} e^{2ni\pi v} \qquad (4.3\text{--}2d)$$

The four theta series (4.3–2) are Laurent expansions in the variable $e^{i\pi v}$ and converge for all finite non-zero values of this variable.

All four theta functions are entire periodic functions of v with the period of θ_1 and θ_2 being 2 and that of θ_3 and θ_4 being 1 (Table 4–1). Table 4–1 also shows their behavior under addition of half and quarter periods using the following abbreviations:

$$A(v) = e^{-i\pi(2v+\tau)} \qquad (4.3\text{--}3a)$$
$$B(v) = e^{-i\pi(v+1/4\tau)} \qquad (4.3\text{--}3b)$$

Table 4–1 shows that all four theta functions may be generated from any one of them by the addition of quarter-periods. From Table 4–1 $\theta_1(v)$ is seen to have a zero at $v = 0$ and thus zeros at $m + n\tau$ where m,n here, as elsewhere in Table 4–1, are integers. It can be proved by integrating θ_1'/θ_1 over the boundary of a parallelogram with vertices $\pm 1/2 \pm 1/2 \tau$ that these are the only zeros of $\theta_1(v)$.

Table 4–1: Some Properties of the Four Theta Functions

	$\theta_1(v)$	$\theta_2(v)$	$\theta_3(v)$	$\theta_4(v)$
zeros	$m+n\tau$	$m+1/2+n\tau$	$m+1/2+(n+1/2)\tau$	$m+(n+1/2)\tau$
$v \to -v$	$-\theta_1(v)$	$\theta_2(v)$	$\theta_3(v)$	$\theta_4(v)$
$v \to v+1$	$-\theta_1(v)$	$-\theta_2(v)$	$\theta_3(v)$	$\theta_4(v)$
$v \to v+\tau$	$-A(v)\theta_1(v)$	$A(v)\theta_2(v)$	$A(v)\theta_3(v)$	$-A(v)\theta_4(v)$
$v \to v+1+\tau$	$A(v)\theta_1(v)$	$-A(v)\theta_2(v)$	$A(v)\theta_3(v)$	$-A(v)\theta_4(v)$
$v \to v+1/2$	$\theta_2(v)$	$-\theta_1(v)$	$\theta_4(v)$	$\theta_3(v)$
$v \to v+1/2\tau$	$iB(v)\theta_4(v)$	$B(v)\theta_3(v)$	$B(v)\theta_2(v)$	$iB(v)\theta_1(v)$
$v \to v+1/2+1/2\tau$	$B(v)\theta_3(v)$	$-iB(v)\theta_4(v)$	$iB(v)\theta_1(v)$	$B(v)\theta_2(v)$

Information on the zeros of the theta functions can be used to obtain infinite products representing the theta functions. Define q_0 as follows:

$$q_0 = \prod_{n=1}^{\infty}(1-q^{2n}) \tag{4.3–4}$$

Then the following infinite products for the theta functions are obtained:

$$\theta_1(v) = 2q_0\sqrt[4]{q}\ \sin \pi v \prod_{n=1}^{\infty}(1-2q^{2n}\cos 2\pi v + q^{4n}) \tag{4.3–5a}$$

$$\theta_2(v) = 2q_0\sqrt[4]{q}\ \cos \pi v \prod_{n=1}^{\infty}(1+2q^{2n}\cos 2\pi v + q^{4n}) \tag{4.3–5b}$$

$$\theta_3(v) = q_0 \prod_{n=1}^{\infty}(1+2q^{2n-1}\cos 2\pi v + q^{4n-2}) \qquad (4.3\text{--}5c)$$

$$\theta_4(v) = q_0 \prod_{n=1}^{\infty}(1-2q^{2n-1}\cos 2\pi v + q^{4n-2}) \qquad (4.3\text{--}5d)$$

These equations (4.3–5) are valid in the entire complex plane representing the complex variable v.

The following relations between squares of theta functions of the same variable are of interest:

$$\theta_1^2(v)\theta_2^2(0) = \theta_4^2(v)\theta_3^2(0) - \theta_3^2(v)\theta_4^2(0) \qquad (4.3\text{--}6a)$$
$$\theta_1^2(v)\theta_3^2(0) = \theta_4^2(v)\theta_2^2(0) - \theta_2^2(v)\theta_4^2(0) \qquad (4.3\text{--}6b)$$
$$\theta_1^2(v)\theta_4^2(0) = \theta_3^2(v)\theta_2^2(0) - \theta_2^2(v)\theta_3^2(0) \qquad (4.3\text{--}6c)$$
$$\theta_4^2(v)\theta_4^2(0) = \theta_3^2(v)\theta_3^2(0) - \theta_2^2(v)\theta_2^2(0) \qquad (4.3\text{--}6d)$$

Each of these equations may be proved by noting that the ratio of its two sides is a doubly periodic function with periods 1 and τ without zeros or poles and is thus a constant. This constant can be evaluated using special values of v (half-periods).

The "theta functions of zero argument" (i.e., $v = 0$ in equations 4.3–1 to 4.3–3) are particularly important and satisfy the following identities:

$$\theta_1'(0) = \pi\theta_2(0)\theta_3(0)\theta_4(0) \qquad (4.3\text{--}7a)$$
$$\theta_2^4(0) + \theta_4^4(0) = \theta_3^4(0) \qquad (4.3\text{--}7b)$$

Graphs illustrating the behavior of theta functions of argument zero are given elsewhere.[7]

Theta functions are very closely related to the Weierstrass sigma functions and can be used for their numerical evaluation because of their extremely rapid convergence.[3] The following equations then express the various Weierstrass functions in terms of theta functions using the notation in equations 4.2–11 to 4.2–14 where $v = z/2\omega$:

$$\sigma(z) = 2\omega \exp\left(\frac{\eta z^2}{2\omega}\right) \frac{\theta_1(v)}{\theta_1'(0)} \qquad (4.3\text{--}8a)$$

---

[7]F. C. Tricomi, *Funzioni Ellittiche*, Zanichelli, Bologna, 1951; German edition, Akad. Verlagsgesellschaft, Leipzig, 1948.

$$\zeta(z) = \frac{\eta z}{\omega} + \frac{\theta_1'(v)}{2\omega\theta_1(v)} \qquad (4.3\text{--}8b)$$

$$\wp(z) = e_a + \frac{1}{4\omega^2}\left[\frac{\theta_1'(0)\theta_{a+1}(v)}{\theta_{a+1}(0)\theta_1(v)}\right]^2 \quad \text{where } a = 1, 2, 3 \qquad (4.3\text{--}8c)$$

$$\wp'(z) = -\frac{\theta_2(v)\theta_3(v)\theta_4(v)\theta_1'^3(0)}{4\omega^3\theta_2(0)\theta_3(0)\theta_4(0)\theta_1^3(v)} \qquad (4.3\text{--}8d)$$

$$12\omega^2 e_1 = \pi^2[\theta_3^4(0) + \theta_4^4(0)] \qquad (4.3\text{--}9a)$$
$$12\omega^2 e_2 = \pi^2[\theta_2^4(0) - \theta_4^4(0)] \qquad (4.3\text{--}9b)$$
$$12\omega^2 e_3 = -\pi^2[\theta_2^4(0) + \theta_3^4(0)] \qquad (4.3\text{--}9c)$$

$$\sqrt{e_2 - e_3} = i\sqrt{e_3 - e_2} = \frac{\pi}{2\omega}\theta_2^2(0) \qquad (4.3\text{--}10a)$$

$$\sqrt{e_1 - e_3} = i\sqrt{e_3 - e_1} = \frac{\pi}{2\omega}\theta_3^2(0) \qquad (4.3\text{--}10b)$$

$$\sqrt{e_1 - e_2} = i\sqrt{e_2 - e_1} = \frac{\pi}{2\omega}\theta_4^2(0) \qquad (4.3\text{--}10c)$$

$$g_2 = \frac{2}{3}\left(\frac{\pi}{2\omega}\right)^4 [\theta_2^8(0) + \theta_3^8(0) + \theta_4^8(0)] \qquad (4.3\text{--}11a)$$

$$g_3 = \frac{4}{27}\left(\frac{\pi}{2\omega}\right)^6 [\theta_2^4(0) + \theta_3^4(0)][\theta_3^4(0) + \theta_4^4(0)][\theta_4^4(0) - \theta_2^4(0)] \qquad (4.3\text{--}11b)$$

$$\sqrt[4]{\Delta} = \frac{\pi}{4\omega^3}\theta_1'^2(0) = \frac{\pi^3}{4\omega^3}[\theta_2(0)\theta_3(0)\theta_4(0)]^2 \qquad (4.3\text{--}12)$$

$$\eta = -\frac{\theta_1'''(0)}{12\omega\theta_1'(0)} \qquad \eta' = -\frac{i\tau\pi\theta_1'''(0)}{24\omega^2\theta_1'(0)} \qquad (4.3\text{--}13)$$

Equation 4.3–8a may be proved by noting that the quotient of the function on its two sides is a doubly periodic function without poles or zeros and approaches 1 as v and z approach 0. Equation 4.3–8b can be obtained by logarithmic differentiation of (4.3–8a).

Elliptic Functions

The Jacobian elliptic functions can also be expressed in terms of theta functions leading to the following equations using the notation in (4.1–9) to (4.1–11):

$$\sqrt{k} = \frac{\theta_2(0)}{\theta_3(0)} \tag{4.3–14a}$$

$$\sqrt{k'} = \frac{\theta_4(0)}{\theta_3(0)} \tag{4.3–14b}$$

$$\text{sn } u = \frac{\theta_3(0)\theta_1(v)}{\theta_2(0)\theta_4(v)} \tag{4.3–15a}$$

$$\text{cn } u = \frac{\theta_4(0)\theta_2(v)}{\theta_2(0)\theta_4(v)} \tag{4.3–15b}$$

$$\text{dn } u = \frac{\theta_4(0)\theta_3(v)}{\theta_3(0)\theta_4(v)} \tag{4.3–15c}$$

Thus given $\tau = \omega'/\omega$, (4.3–14a) determines the modulus k of the Jacobian elliptic functions and equations 4.3–15 determine the functions themselves.

Many applications of elliptic functions require calculation of a q when $|q| < 1$ from a given k^2 defined by

$$k^2 = \frac{\theta_2^4(0,q)}{\theta_3^4(0,q)} = 1 - \frac{\theta_4^4(0,q)}{\theta_3^4(0,q)} = 1 - (k')^2 \tag{4.3–16}.$$

This is called the *inversion problem*. Frequently $0 < k^2 < 1$ and

$$\frac{\theta_4^4(0,q)}{\theta_3^4(0,q)} = \prod_{n=1}^{\infty} \frac{(1 - q^{2n-1})}{(1 + q^{2n+1})} \tag{4.3–17}$$

by equations 4.3–5. As q increases from 0 through real values to 1, the infinite product decreases monotonically from 1 to 0 so that (4.3–16) has exactly one solution q in the range $0 < q < 1$.

The inversion problem thus involves the calculation of a value of the theta function parameter q corresponding to a set of roots e_1, e_2, and e_3 of the cubic equation 4.2–17 = 4.2–18. The first step involves calculation of the parameter L defined by the following equation:

$$L = \frac{1 - \sqrt{k'}}{1 + \sqrt{k'}} = \frac{\sqrt[4]{e_1 - e_3} - \sqrt[4]{e_1 - e_2}}{\sqrt[4]{e_1 - e_3} + \sqrt[4]{e_1 - e_2}} \qquad (4.3\text{–}18).$$

The corresponding value of q can then be obtained by inversion of the equation[8]

$$\frac{L}{2} = \frac{q + q^9 + q^{25} + \cdots}{1 + 2q^4 + 2q^{16} + \cdots} \qquad (4.3\text{–}19)$$

to give the so-called *Jacobi nome*

$$q = \left(\frac{L}{2}\right) + 2\left(\frac{L}{2}\right)^5 + 12\left(\frac{L}{2}\right)^9 + 150\left(\frac{L}{2}\right)^{13} + \cdots = \sum_{j=1}^{\infty} q_j \left(\frac{L}{2}\right)^{4j+1} \qquad (4.3\text{–}20)$$

in which the coefficients q_j form the series[7]

1; 2; 15; 150; 1707; 20,910; 268,616; 3,567,400; 48,555,069; 673,458,874; 9,481,557,398; 135,119,529,972; 1,944,997,539,623; 28,235,172,753,886

Equation 4.3–18 poses some difficulty since for a given ordering of the roots e_1, e_2, e_3 of the cubic equation 4.2–17 there are four fourth roots $\sqrt[4]{e_1 - e_3}$ and four fourth roots $\sqrt[4]{e_1 - e_2}$. The 16 possible combinations of these two fourth roots reduce to four possible values of L when the quotient (4.3–18) is taken. In addition, there are six permutations of the three roots e_1, e_2, and e_3 leading to (4)(6) = 24 possible q values for a given cubic equation 4.2–17. In general, half of these possible q values will have $|q| > 1$ and

---

[8] A. N. Lowan, G. Blanch, and W. Horenstein, On the Inversion of the q-Series Associated with Jacobian Elliptic Functions, *Bull. Amer. Math. Soc.*, **48**, 737–738 (1942).

therefore are unsatisfactory since the series for the corresponding theta functions do not converge. Early work by Weierstrass[4] suggests the following criteria for choosing the correct ordering of e_1, e_2, e_3 and the correct fourth roots in equation 4.3–18:

(1) The labelling e_1, e_2, e_3 of the roots of equation 4.2–17 is chosen so that neither the expression 4.2–21c for k^2 nor 4.2–21d for k'^2 has a negative real value and also neither of the two expressions has a positive real value which is equal to or greater than one;

(2) The initial square roots required to convert the expressions 4.2–21c for k^2 and 4.2–21d for k'^2 to k and k', respectively, are taken so that the real parts are positive leading to a positive real part for the quotient $\frac{k'}{k}$;

(3) The fourth roots are taken so that the quotients

$$\sqrt{k} = \frac{\sqrt[4]{e_2 - e_3}}{\sqrt[4]{e_1 - e_3}} = \sqrt[4]{\frac{e_2 - e_3}{e_1 - e_3}} \qquad (4.3\text{–}21a)$$

$$\sqrt{k'} = \frac{\sqrt[4]{e_1 - e_2}}{\sqrt[4]{e_1 - e_3}} = \sqrt[4]{\frac{e_1 - e_2}{e_1 - e_3}} \qquad (4.3\text{–}21b)$$

have the principal values.

4.4 Higher Order Theta Functions

The definition of elliptic integrals by an equation of the type

$$f(x) = \int \frac{dx}{\sqrt{\mathcal{P}(x)}} \qquad (4.4\text{–}1)$$

where $\mathcal{P}(x)$ is a cubic or quartic polynomial can be extended to hyperelliptic integrals where $\mathcal{P}(x)$ is a polynomial of degree 5 or higher. The corresponding higher order theta functions are required for solutions of general algebraic equations of degree 6 or higher.

The simple theta functions discussed in the previous section may be expressed as singly infinite series of exponentials of polynomials which are quadratic in the summation index, e.g., from (4.3–2c),

$$\theta_3(v) = \sum_{n=-\infty}^{\infty} q^{n^2} e^{2ni\pi v} \qquad (4.4-2)$$

Since $q = e^{i\pi\tau}$ such functions can be expressed generally in the form

$$\theta(v) = \sum_{n=-\infty}^{\infty} \exp(2ni\pi v + i\pi\tau n^2) \qquad (4.4-3)$$

where τ is a complex constant whose imaginary part is positive to ensure convergence.[9] There are three other functions functions $\theta_{\alpha\beta}$ connected with this one and obtained from by replacing n by $n + 1/2\alpha$ and v by $v + 1/2\beta$ where α and β may be 0 or 1. The four single theta functions are given the labels $\theta_1(v)$, $\theta_2(v)$, $\theta_3(v)$, and $\theta_4(v)$ in the previous section (e.g., equations 4.3–2). Various ratios of these single theta functions are elliptic functions (e.g., equations 4.3–8).

The concept of theta functions can be generalized to a multiply infinite series with several arguments in which τ is now a symmetric square matrix rather than a single complex number. The *genus* of such a theta function is the size of its τ matrix. The genus 2 or *double* theta functions are important in the solution of the general sextic equation and can be written as follows:

$$\theta_{\alpha\beta}(v) = \sum\sum \exp[2\pi i(n+1/2\alpha)(v+1/2\beta) + \pi i\tau(n+1/2\alpha)^2] \qquad (4.4-4)$$

in which τ is a 2×2 symmetric matrix, i.e.,

$$\tau = \begin{pmatrix} \tau_{11} & \tau_{12} \\ \tau_{21} & \tau_{22} \end{pmatrix} \text{ and } \tau_{12} = \tau_{21} \qquad (4.4-5)$$

and all other letters are row designations standing for pairs of letters

---

[9] R. W. H. T. Hudson, *Kummer's Quartic Surface*, Cambridge, 1905, Chapter XVI.

distinguished by suffixes 1 and 2. The summation is taken over all integer values of n_1 and n_2 ranging from $-\infty$ to $+\infty$. The variables α_1 and α_2 are integers which are taken to be 0 or 1 since the integer parts of $1/2\alpha$ may be absorbed in n, β_1 and β_2 are also integers, and since the addition of even integers to β can at most change the sign of the function, it will generally be supposed that β_1 and β_2 are either 0 or 1. The matrix $\begin{pmatrix} \alpha_1 \alpha_2 \\ \beta_1 \beta_2 \end{pmatrix}$ is called the *characteristic* of the double theta function and can be indicated by the suffix $\alpha\beta$. The designation $\alpha\beta$ denotes $\alpha_1\beta_1 + \alpha_2\beta_2$, and the parity of the theta function depends upon the parity of this expression since by taking a new pair of summation letters $n' = -n - \alpha$, the order of the terms is changed without altering the value of the function.

There are $2^{2^2} = 16$ different characteristics for the double theta functions since each of the four elements α_1, α_2, β_1, and β_2 can be 0 or 1. The double theta function in which all of the elements are 0 is called the *zero characteristic*. Corresponding to these double theta functions of zero characteristic are 16 *half periods* $1/2(\tau\alpha + \beta)$. The effect of adding a half period to the argument by a theta function is to change the characteristic and multiply by a nonvanishing function. Of the 16 characteristics six are odd and ten are even. Hence six double theta functions are odd functions and vanish for $v = 0$.

Chapter 5

Algebraic Equations Soluble by Radicals

5.1 The Quadratic and Cubic Equations

The general monic quadratic equation can be written as
$$x^2 + a_1 x + a_2 = 0 \tag{5.1-1}.$$
The Galois group of this equation, $S_2 \approx C_2$, has only the two elements $\{1, (12)\}$ and the corresponding alternating group $A_2 \approx C_1$ contains only the identity element. The trivial normal series can be described as $G_0 = C_1 = A_1 \triangleleft C_2 = S_2$ with a factor group $S_2/A_2 = C_2$. The two roots of the equation are obtained by the following well-known quadratic formula:

$$x_k = \frac{-a_1 \pm \sqrt{a_1^2 - 4a_2}}{2} \qquad k = 1,2 \tag{5.1-2}$$

The C_2 factor group corresponds to the square root in equation 5.1–2. The nonidentity permutation of the $S_2 = C_2$ Galois group of this equation, namely (12), is seen to permute the two roots by changing the sign of the radical. The quantity under the radical in (5.1–2), namely $a_1^2 - 4a_2$, is called the *discriminant*, D, of the quadratic equation. If $D = 0$, the two roots of the quadratic equation are equal, namely $x_1 = x_2$. If $D > 0$, the quadratic equation 5.1–1 has two distinct real roots whereas if $D < 0$ the quadratic equation has two distinct nonreal complex roots which necessarily are complex conjugates.

The solution of the general monic cubic equation[1]
$$x^3 + a_1 x^2 + a_2 x + a_3 = 0 \tag{5.1-3}$$
is much more interesting. The Galois group of this equation, S_3, with $3! = 6$ elements, is isomorphic with the symmetry group of the equilateral triangle, namely D_{3h}. The S_3 group has the following normal series and associated factor groups:

---

[1] E. Dehn, *Algebraic Equations*, Columbia University Press, New York, 1930.

Equations Soluble by Radicals

$$G_0 = C_1 \triangleleft C_3 = A_3 \triangleleft S_3 \tag{5.1-4}$$

$$A_3/C_1 = C_3 \tag{5.1-5a}$$

$$S_3/A_3 = C_2 \tag{5.1-5b}$$

The solution of such an equation based on this normal series is thus seen to involve a square root from the factor group $S_3/A_3 = C_2$ inside the cube root from the factor group $A_3/C_1 = C_3$.

In the actual algorithm to solve such a cubic equation (5.1-3)[2] the equation is first simplified by applying a Tschirnhausen transformation (Section 3.3) to the general cubic equation (5.1-3) to give the following reduced cubic equation with no term of degree 2:

$$z^3 + b_2 z + b_3 = 0 \tag{5.1-6}$$

where

$$x = z - \frac{a_1}{3} \tag{5.1-7a}$$

$$b_2 = \frac{3a_2 - a_1^2}{3} \tag{5.1-7b}$$

$$b_3 = \frac{2a_1^3 - 9a_1 a_2 + 27 a_3}{27} \tag{5.1-7c}$$

The remainder of the algorithm consists of methods for determination of the roots of the reduced cubic equation (5.1-6). In this connection we seek to determine two numbers u and v in such a manner that

$$z = u + v \tag{5.1-8}$$

will satisfy equation (5.1-6). Substitution $u + v$ for z in equation (5.1-6) gives the following equation:

$$u^3 + v^3 + 3(uv + b_2/3)(u + v) + b_3 = 0 \tag{5.1-9}$$

Equation (5.1-9) will be satisfied if u and v are chosen in accord with the following relationships:

$$u^3 + v^3 = b_3 \tag{5.1-10a}$$

$$uv = \frac{-b_2}{3} \tag{5.1-10b}$$

Solving equations (5.1-10) simultaneously gives

$$u^6 + b_3 u^3 - \frac{b_3^3}{27} = 0 \tag{5.1-11}$$

Solving equation 5.1-11 as a quadratic equation in u^3 gives

---

[2] N. B. Conkwright, *Introduction to the Theory of Equations*, Ginn, Boston, 1941, Chapter 5.

$$u^3 = \frac{-b_3 \pm \sqrt{b_3^2 + \frac{4b_2^3}{27}}}{2} = \frac{-b_3}{2} \pm \sqrt{\frac{b_3^2}{4} + \frac{b_2^3}{27}} \qquad (5.1\text{--}12)$$

If u_1 denotes any of the cube roots of the number on the right hand side of equation (5.1–2), the three cube roots are u_1, ωu_1, and $\omega^2 u_1$ where $\omega = \exp(2\pi i/3)$ (i.e., a complex cube root of unity). The corresponding values of v, obtained from equation (5.1–10b), are $-\frac{b_2}{3u_1}$, $-\frac{b_2}{3\omega u_1}$, and $-\frac{b_2}{3\omega^2 u_1}$ or $-\frac{b_2}{3u_1}$, $-\frac{\omega^2 b_2}{3u_1}$, and $-\frac{\omega b_2}{3u_1}$, respectively, since $\omega^3 = 1$. Now if $-\frac{b_2}{3u_1}$ is denoted by v_1, then the three roots of the reduced cubic equation (5.1–6) may be written in the following form:

$$z_1 = u_1 + v_1 \qquad (5.1\text{--}13a)$$
$$z_2 = \omega u_1 + \omega^2 v_1 \qquad (5.1\text{--}13b)$$
$$z_3 = \omega^2 u_1 + \omega v_1 \qquad (5.1\text{--}13c)$$

Combining equations (5.1–12) and (5.1–13) leads to the following relations, called *Cardan's formulas*, for the roots z_k ($k = 1, 2, 3$) of the reduced cubic equation (5.1–6):

$$z_k = \sqrt[3]{\frac{-b_3}{2} + \sqrt{\frac{b_3^2}{4} + \frac{b_2^3}{27}}} + \sqrt[3]{\frac{-b_3}{2} - \sqrt{\frac{b_3^2}{4} + \frac{b_2^3}{27}}} \qquad (5.1\text{--}14)$$

In (5.1–14) the cube roots for the two terms are selected so that their product is always $-b_2/3$ in accord with equation (5.1–10b). This solution of the general cubic equation (5.1–3) thus relates to the factorization of its S_3 Galois group with the C_3 factor (5.1–5a) corresponding to the outer cube roots in equation 5.1–14 and the C_2 factor (5.1–5b) corresponding to the inner square roots in (5.1–14) arising from solution of the quadratic equation (5.1–12) in u^3.

The cubic equation, like the quadratic equation, has a discriminant that determines the nature of its roots. The discriminant of the general cubic equation 5.1–3 is essentially the quantity under the *square* root signs in equation 5.1–14, which is conveniently written as

$$\Delta = 4b_2^3 + 27b_3^2 \qquad (5.1\text{--}15)$$

Equations Soluble by Radicals

The following three situations cover all possibilities for the three roots of a general cubic equation (5.1–3):
(1) $\Delta = 0 \Rightarrow$ there is a multiple root;
(2) $\Delta > 0 \Rightarrow$ there is exactly one real root and two nonreal complex conjugate roots;
(3) $\Delta < 0 \Rightarrow$ the three roots are real and distinct.

The case of three real and distinct roots is historically called the *casus irreducibilis* (irreducible case),[3] since evaluation of the roots of such equations using Cardan's formulas (equation 5.1–14) requires taking real cube roots of nonreal numbers. Thus, although the roots of such cubic equations can be expressed using only radicals in addition to simple arithmetical operations (addition, subtraction, multiplication, and division), evaluation of these radicals in these expressions is not feasible. This difficulty is circumvented by using trigonometric functions rather than radicals to solve such cubic equations with three real roots.

The radicals in Cardan's formulas (equation 5.1–14) may be evaluated using trigonometric functions by first using the relationships:

$$\sqrt[3]{\frac{-b_3}{2} + \sqrt{\frac{b_3^2}{4} + \frac{b_2^3}{27}}} = \sqrt[3]{\frac{-b_3}{2} + i\sqrt{\frac{-b_3^2}{4} + \frac{-b_2^3}{27}}} \qquad (5.1\text{–}16\text{a})$$

$$\sqrt[3]{\frac{-b_3}{2} - \sqrt{\frac{b_3^2}{4} + \frac{b_2^3}{27}}} = \sqrt[3]{\frac{-b_3}{2} - i\sqrt{\frac{-b_3^2}{4} + \frac{-b_2^3}{27}}} \qquad (5.1\text{–}16\text{b})$$

The absolute value of the complex number on the right hand side of equation (5.1–16a) is $\sqrt{\frac{-b_2^3}{27}}$, and the amplitude may be found by the following relation:

$$\cos \phi = \frac{-b_3}{2\sqrt{\frac{-b_2^3}{27}}} \qquad (5.1\text{–}17)$$

---

[3] D. E. Dobbs and R. Hanks, *A Modern Course on the Theory of Equations*, Polygonal Publishing House, Passaic, NJ, 1980, Chapter 3.

The angle ϕ terminates in the first quadrant (i.e., $0° < \phi < 90°$) if b_3 is negative (i.e., $\cos \phi > 0$) and in the second quadrant (i.e., $90° < \phi < 180°$) if b_3 is positive (i.e., $\cos \phi < 0$). This leads to the following expressions for the radicals in Cardan's formulas in terms of trigonometric functions:

$$\frac{-b_3}{2} + \sqrt{\frac{b_3{}^2}{4} + \frac{b_2{}^3}{27}} = \sqrt{\frac{-b_2{}^3}{3}}(\cos \phi + i \sin \phi) \qquad (5.1\text{-}18\text{a})$$

$$\frac{-b_3}{2} - \sqrt{\frac{b_3{}^2}{4} + \frac{b_2{}^3}{27}} = \sqrt{\frac{-b_2{}^3}{3}}(\cos \phi - i \sin \phi) \qquad (5.1\text{-}18\text{b})$$

Taking the cube roots of these equations gives the following where $k = 1, 2, 3$:

$$\sqrt[3]{\frac{-b_3}{2} + \sqrt{\frac{b_3{}^2}{4} + \frac{b_2{}^3}{27}}} = \sqrt{\frac{-b_2}{3}}\left(\cos \frac{\phi + 2k\pi}{3} + i \sin \frac{\phi + 2k\pi}{3}\right) \qquad (5.1\text{-}19\text{a})$$

$$\sqrt[3]{\frac{-b_3}{2} - \sqrt{\frac{b_3{}^2}{4} + \frac{b_2{}^3}{27}}} = \sqrt{\frac{-b_2}{3}}\left(\cos \frac{\phi + 2k\pi}{3} - i \sin \frac{\phi + 2k\pi}{3}\right) \qquad (5.1\text{-}19\text{b})$$

These equations lead to the following formulas for the roots of the reduced cubic equation (5.1–6) in terms of trigonometric functions:

$$\phi = \cos^{-1}\left(\frac{-b_3}{2\sqrt{\frac{-b_2{}^3}{27}}}\right) \qquad (5.1\text{-}20\text{a})$$

$$z_k = 2\sqrt{\frac{-b_2}{3}} \cos\left(\frac{\phi + 2k\pi}{3}\right) \qquad k = 1, 2, 3 \qquad (5.1\text{-}20\text{b})$$

In equations 5.1–20 the C_3 factor group corresponds to the division of $\phi + 2k\pi$ by 3 before taking the cosine (5.1–20b) whereas the C_2 factor group corresponds to the square roots in the formulas. Note that the sequence of steps 5.1–20a and 5.1–20b first involves taking the inverse cosine of one number followed by taking the cosine of another number derived from the first inverse

Equations Soluble by Radicals

cosine after dividing by 3. This procedure is analogous to evaluating a cube root by taking its logarithm, dividing the logarithm by 3, and then taking the inverse logarithm of the quotient. Both the logarithms needed in general to evaluate radicals (e.g., the square and cube roots in equation 5.1–14) and the inverse cosine needed for equation 5.1–20a can be expressed as integrals of the type (compare equation 4.1–3 in Section 4.1)

$$f(y) = \int \frac{dy}{\sqrt{\mathcal{P}(y)}} \tag{5.1-21}$$

in which $\mathcal{P}(y)$ is a *quadratic* polynomial, i.e., y^2 for $\ln a$ and $1-y^2$ for $-\cos^{-1}a$. This makes readily apparent the close relationship between the two methods for solving cubic equations, namely that represented by equation 5.1–14 and that represented by equations 5.1–20.

5.2 The Quartic Equation

Increasing the degree of the algebraic equation from 3 to 4 introduces no fundamentally new features. The Galois group of the general quartic, S_4, with $4! = 24$ elements, is isomorphic to the symmetry point group of the regular tetrahedron T_d (Section 2.3). The S_4 group has the following normal series and associated factor groups:

$$G_0 = C_1 \triangleleft C_2 \triangleleft D_2 \triangleleft A_4 \triangleleft S_4 \approx T_d \tag{5.2-1}$$

$$C_2/C_1 = C_2 \tag{5.2-2a}$$

$$D_2/C_2 = C_2 \tag{5.2-2b}$$

$$\left.\begin{array}{l} A_4/D_2 = C_3 \\ S_4/A_4 = C_2 \end{array}\right\} \text{cubic equation (see section 5.1)} \quad \begin{array}{l}(5.2-2c)\\(5.2-2d)\end{array}$$

Thus the solution of a quartic equation can be reduced to the solution of a cubic equation (called the resolvent cubic equation) corresponding to the C_3 and C_2 factor groups in equations 5.2–2c and 5.2–2d, respectively, followed by solutions of two additional quadratic equations corresponding to equations 5.2–2a and 5.2–2b.

A classical method for solving the quartic equation which has these features is Ferrari's solution of the general quartic equation.[2] Consider the following general monic quartic equation:

$$x^4 + a_1x^3 + a_2x^2 + a_3x + a_4 = 0 \qquad (5.2\text{--}3)$$

Determine the numbers a, b, and k such that

$$x^4 + a_1x^3 + a_2x^2 + a_3x + a_4 + (ax+b)^2 \equiv \left(x^2 + \frac{a_1}{2}x + k\right)^2 \qquad (5.2\text{--}4)$$

by equating the coefficients of like powers of x on both sides of this equation leading to the following relations:

From the x^2 term: $\qquad a^2 + a_2 = 2k + \dfrac{a_1^2}{4} \qquad (5.2\text{--}5a)$

From the x term: $\qquad 2ab + a_3 = ka_1 \qquad (5.2\text{--}5b)$

From the constant term: $\qquad b^2 + a_4 = k^2 \qquad (5.2\text{--}5c)$

Solving the three simultaneous equations for a, b, and c leads to the following cubic equation for k, known as the *resolvent cubic*:

$$k^3 - \frac{a_2}{2}k^2 + \frac{1}{4}(a_1a_3 - 4a_4)k + \frac{1}{8}(4a_2a_4 - a_1^2a_4 - a_3^2) = 0 \qquad (5.2\text{--}6)$$

Solving this cubic equation (5.2–6) for k using either (5.1–14) (the radical method) or equations 5.1–20 (the trigonometric method) followed by substitution in (5.2–5a) and (5.2–5b) gives a and b. Combining equations 5.2–3 and 5.2–4 gives the following equation in which both sides are perfect squares:

$$(x^2 + \frac{a_1}{2}x + k)^2 = (ax+b)^2 \qquad (5.2\text{--}7)$$

Taking one square root gives the following two quadratic equations:

$$x^2 + \frac{a_1}{2}x + k = ax + b \qquad (5.2\text{--}8a)$$

$$x^2 + \frac{a_1}{2}x + k = -ax - b \qquad (5.2\text{--}8b)$$

The four roots of the original quartic equation 5.2–3 are then obtained pairwise from solutions of the two quadratic equations 5.2–8.

The Ferrari method for solving the general quartic equation 5.2–3 corresponds to the factor group structure 5.2–1 and 5.2–2 for the symmetric group $S_4 \approx T_d$ as follows:

(1) $A_4/D_2 = C_3$ and $S_4/A_4 = C_2$ correspond to the resolvent cubic equation 5.2–6;

(2) $D_2/C_2 = C_2$ corresponds to taking the square roots of both sides of equation 5.2-7;

(3) $C_2/C_1 = C_2$ corresponds to solving the quadratic equations 5.2-8.

5.3 Special Quintic Equations Solvable by Radicals

Consider an irreducible monic quintic equation, i.e.
$$x^5 + a_1 x^4 + a_2 x^3 + a_3 x^2 + a_4 x + a_5 = 0 \qquad (5.3\text{-}1).$$
Its Galois group must be a transitive permutation group for the five roots x_k ($1 \le k \le 5$). Table 2-3 indicates that there are five different transitive permutation groups of degree 5, namely $C_5, D_5, M_5, A_5 \equiv I$, and S_5 with 5, 10, 20, 60, and 120 operations, respectively. Among these five groups the first three, namely C_5, D_5, and M_5, are soluble, whereas the remaining two groups, namely A_5 and S_5 are not soluble. Thus irreducible quintic equations with C_5, D_5, and M_5 Galois groups are soluble by radicals whereas the more general irreducible quintic equations with A_5 and S_5 are not soluble by radicals. Berwick[4] has characterized quintic equations with these Galois groups and thus soluble by radicals.

The metacyclic group M_5 contains the 20 permutations of the general type $s^m t^n$ ($m = 1, 2, 3, 4, 5$; $n = 1, 2, 3, 4$) where $s = (12345)$ and $t = (1234)$ so that $s^5 = t^4 = 1$ and $st = ts^2$ (or $t^{-1}st = s^2$). The dihedral group D_5 is a normal subgroup of M_5 of index 2 containing the 10 permutations of the general type $s^m u^n$ ($m = 1, 2, 3, 4, 5$; $n = 1, 2$) where $u = t^2 = (12)(34)$ and $s^5 = u^2 = 1$ and $su = us^4$ (or $u^{-1}su = s^4$). The cyclic group C_5 consists of the single cycle $s, s^2, s^3, s^4, s^5 \equiv 1$. The cyclic group C_5 consists of the fivefold proper rotations of the regular pentagon, i.e., $C_5{}^n$ ($n = 1, 2, 3, 4$) and $C_5{}^5 \equiv E$ whereas the dihedral group D_5 consists of not only the fivefold proper rotations of the pentagon within its plane but also its twofold proper rotations ("flips") out of its plane (Figure 5-1). The metacyclic group M_5 contains not only all of the proper rotations of the regular pentagon but the proper rotations that can convert a regular pentagon into the pentagram having the same vertices as the

---

[4]W. E. H. Berwick, The Condition that a Quintic Equation Should be Soluble by Radicals, *Proc. London Math. Soc.*, (2) **14**, 301 (1915).

original pentagon (Figure 5–1). Furthermore the group M_5 has the following normal series and associated factor groups:

$$G_0 = C_1 \triangleleft C_5 \triangleleft D_5 \triangleleft M_5 \qquad (5.3\text{–}2)$$

$$C_5/C_1 = C_5 \qquad (5.3\text{–}3a)$$

$$D_5/C_5 = C_2 \qquad (5.3\text{–}3b)$$

$$M_5/D_5 = C_2 \qquad (5.3\text{–}3c)$$

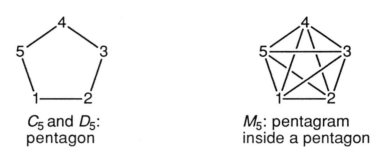

C_5 and D_5: pentagon

M_5: pentagram inside a pentagon

Figure 5–1: The regular pentagon and the pentagram inside a pentagon and their relationships to the soluble transitive permutation groups of degree 5, C_5, D_5, and M_5. The numbers on the vertices correspond to the indices k on the roots x_k of the corresponding quintic.

Consider the problem of solving a monic quintic equation 5.3–1 by radicals in a field of rationality K which contains the coefficients a_n ($1 \le n \le 5$). For the existence of such a radical solution of the quintic it is necessary and sufficient that some rational function of the roots x_k ($1 \le k \le 5$), which is unaltered by the 20 permutations of k in M_5 but no others, should have a value rational in the field K.

The quintic 5.3–1 can be reduced to the Bring-Jerrard form

$$y^5 - A_4 y + A_5 = 0 \qquad (5.3\text{–}4)$$

by a Tschirnhausen transformation requiring solution of a cubic equation as well as a quadratic equation (Section 3.3). In general the coefficients A_4 and A_5 of equation 5.3–4 belong to a larger field L where the extension $L:K$ (Section 3.1) is of degree 6 because of the nature of this Tschirnhausen transformation. The coefficients A_4 and A_5 are rational in the original field K only for a certain

Equations Soluble by Radicals

limited class of quintics. Runge[5] defines a function ψ of the five roots x_k of the quintic:

$$\psi = 1/4(x_1x_2 + x_2x_3 + x_3x_4 + x_4x_5 + x_5x_1 - x_1x_3 - x_3x_5 - x_5x_2 - x_2x_4 - x_4x_1)^2$$
(5.3–5)

in which the products $x_j x_k$ corresponding to the edges of the pentagon and pentagram in Figure 5–1 have positive and negative signs, respectively. He then shows that ψ is a root of the following sextic equation:

$$\Psi(\psi) \equiv (\psi - A_4)^4((\psi^2 - 6\psi A_4 + 25A_4^2) - 5^5 A_5^4 \psi = 0 \quad (5.3\text{–}6)$$

If (5.3–6) has a rational root in the field L, the quintic can be solved by radicals in L and consequently in K. Thus the problem of identifying quintic equations soluble by radicals consists of obtaining a sextic equation $\Phi(\phi) = 0$ satisfied by a function of symmetry M_5 but with rational coefficients in K. Since two rational functions of the roots which belong to the same group are rational functions of each other, the sextic equation is connected with equation 5.3–4 by a transformation of one of the following types where y_k and z_k are rational in L:

$$\phi = y_0 + y_1\psi + y_2\psi^2 + y_3\psi^3 + y_4\psi^4 + y_5\psi^5 \quad (5.3\text{–}7a)$$
$$\psi = z_0 + z_1\phi + z_2\phi^2 + z_3\phi^3 + z_4\phi^4 + z_5\phi^5 \quad (5.3\text{–}7b)$$

The discovery of roots of the sextic equation $\Phi(\phi) = 0$ rational in K is a simpler process that that of roots of $\Psi(\psi) = 0$ rational in L. Thus if the coefficients of the original quintic equation 5.3–1, a_n ($1 \leq n \leq 5$), are ordinary rational numbers, the extended field L is in general a soluble sextic field, and the testing of a sextic equation whose coefficients belong to a sextic field for roots rational in that field, although possible in a finite number of steps, is a somewhat lengthy operation.

The following function of the roots of the quintic equation 5.3–1 can be shown to be unaltered by the permutations of the metacyclic group M_5:

$$\phi = (x_1 - x_2)^2(x_2 - x_3)^2(x_3 - x_4)^2(x_4 - x_5)^2(x_5 - x_1)^2$$
$$+ (x_1 - x_3)^2(x_3 - x_5)^2(x_5 - x_2)^2(x_2 - x_4)^2(x_4 - x_1)^2 \quad (5.3\text{–}8).$$

In equation 5.3–8 the first and second terms correspond to the edges of the pentagon and pentagram, respectively, in Figure 5–1. The number ϕ is one of a

---

[5] C. Runge, Über die auflösbaren Gleichungen von der form $x^5 + ux + v = 0$, Acta Math., 7, 173–186 (1886).

set of six conjugates which may be obtained by the following six permutations of the indices k of the roots x_k:

$$1, (123), (132), (23), (31), (12) \tag{5.3-9}$$

These permutations on the first three indices 1,2,3 and leaving the indices 4 and 5 unchanged form the dihedral group D_3. The group of each of the conjugate functions is a subgroup (although, of course, not a normal subgroup) of the symmetric group S_5 conjugate to M_5. A symmetric function of the six ϕ's is an invariant[6] of the binary homogeneous quintic form

$$f(x,y) \equiv x^5 + a_1 x^4 y + a_2 x^3 y^2 + a_3 x^2 y^3 + a_4 xy^4 + a_5 y^5 \tag{5.3-10}$$

and the six elementary symmetric functions are rational integral invariants. Thus the six ϕ's can be assumed to be roots of a sextic equation

$$\Phi(\phi) \equiv \phi^6 + A_4 \phi^5 + A_8 \phi^4 + A_{12} \phi^3 + A_{16} \phi^2 + A_{20} \phi + A_{24} = 0 \tag{5.3-11}$$

in which $A_4, A_8, A_{12}, A_{16}, A_{20}$, and A_{24} are rational integral invariants of degrees 4, 8, 12, 16, 20, and 24, respectively.

The methods of invariant theory (Section 2.6)[6] can be used to evaluate the coefficients $A_4, A_8, A_{12}, A_{16}, A_{20}$, and A_{24} of equation 5.3–11. The quintic form $f(x,y)$ in equation 5.3–10 has the following four irreducible invariants of which the first three are needed:

$$\text{degree } 4: J = D = (i,i)^2 \tag{5.3-12a}$$

$$\text{degree } 8: K = (i^3, H)^6 \tag{5.3-12b}$$

$$\text{degree } 12: L = (i^5, f^2)^{10} \tag{5.3-12c}$$

$$\text{degree } 18: I = (i^7, fT)^{14} \tag{5.3-12d}$$

In equations 5.3–12 i is the fourth transvectant $(f,f)^4$ and thus of degree 2 and order 2, and D is the discriminant of i, which is obtained by the Hessian $(i,i)^2$. Furthermore, H is the Hessian of f, namely $(f,f)^2$, and thus of degree 6 and order 2, and T is the functional determinant of f and H, namely (f,H) and thus of degree 9 and order 3. The value of i (also called S) can be obtained as follows:

$$i = 1/100[(20a_4 - 8a_1 a_3 + 3a_2^2)x^2 + (100a_5 - 12a_1 a_4 + 2a_2 a_3)xy$$
$$+ (20a_1 a_5 - 8a_2 a_4 + 3a_3^2)y^2]$$
$$= Ax^2 + Bxy + Cy^2 \tag{5.3-13}$$

---

[6] O. E. Glenn, *A Treatise on the Theory of Invariants*, Ginn and Co., Boston, 1915, pp. 148–150.

Equations Soluble by Radicals

Then $J = D$ is the discriminant of the quadratic form 5.3–13, namely
$$J = B^2 - 4AC \qquad (5.3\text{–}14)$$
The functional determinant T can be obtained from the expression

$$T \equiv \frac{1}{1000} \begin{vmatrix} 10x+2a_1y & 2a_1x+a_2y & a_2x+a_3y \\ 2a_1x+a_2y & a_2x+a_3y & a_3x+2a_4y \\ a_2x+a_3y & a_3x+2a_4y & 2a_4x+10a_5y \end{vmatrix}$$

$$\equiv Dx^3 + Ex^2y + Fxy^2 + Gy^3 \qquad (5.3\text{–}15)$$

Then K and L are given by the following equations in which $A, B, C, D, E, F,$ and G are defined as in equations 5.3–14 and 5.3–15:
$$K = 2A(3EG - F^2) - B(9DG - EF) + 2C(3FD - E^2) \qquad (5.3\text{–}16a)$$
$$L = \tfrac{1}{3}[4(3EG - F^2)(3FD - E^2) - (9DG - EF)^2] \qquad (5.3\text{–}16b)$$

The rational equation 5.3–11 for ϕ can now be written as follows by combining J, K, and L (equations 5.3–14, 5.3–16a, and 5.3–16b) in all possible ways to get invariants of the required degrees 4, 8, 12, 16, 20, and 24 for the coefficients $A_4, A_8, A_{12}, A_{16}, A_{20},$ and A_{24}, respectively:

$$\begin{aligned}\Phi(\phi) \equiv\ & \phi^6 + b_1 J \phi^5 + (b_2 J^2 + b_3 K)\phi^4 + (b_4 J^3 + b_5 JK + b_6 L)\phi^3 \\ & + (b_7 J^4 + b_8 J^2 K + b_9 K^2 + b_{10} JL)\phi^2 \\ & + (b_{11} J^5 + b_{12} J^3 K + b_{13} JK^2 + b_{14} J^2 L + b_{15} KL)\phi \\ & + b_{16} J^6 + b_{17} J^4 K + b_{18} J^2 K^2 + b_{19} K^3 + b_{20} J^3 L + b_{21} JKL \\ & + b_{22} L^2 = 0 \end{aligned} \qquad (5.3\text{–}17)$$

The numerical coefficients $b_1, b_2, \ldots, b_{22}$ can be found by taking a sufficient number of special quintics, whether irreducible or not, and comparing the actual coefficients of (5.3–17) with those involving the b's. Berwick[4] used the quintics $x^5 - n = 0$, $x^5 - x = 0$, and quintics with the roots $\{1,-1,3,-3,0\}$, $\{1,1,-1,-1,0\}$, $\{1,2,-2,0,0\}$, $\{1,3,-3,0,0\}$, $\{1,-1,\rho,\rho^2,0\}$, and $\{1,-1,2,-2,0\}$ to evaluate these b's where ρ is given by the equation $\rho^2 + \rho + 1 = 0 \Rightarrow \rho = \tfrac{1}{2}(-1 \pm \sqrt{-3})$. The numerical coefficients in the final results can be simplified by defining the following parameters:

$$j = 5^3 J \qquad (5.3\text{--}18\text{a})$$
$$k = 2^5 5^6 K \qquad (5.3\text{--}18\text{b})$$
$$l = -2^{10} 5^9 L \qquad (5.3\text{--}18\text{c})$$

Then equation 5.3–17 becomes

$$\Phi(\phi) \equiv \phi^6 + 10j\phi^5 + (35j^2+10k)\phi^4 + (60j^3 + 30jk + 10l)\phi^3$$
$$+ (55j^4 + 30j^2k + 25k^2 + 50jl)\phi^2$$
$$+ (26j^5 + 10j^3k + 44jk^2 + 59j^2l + 14kl)\phi$$
$$+ 5j^6 + 20j^2k^2 + 20j^3l + 20jkl + 25l^2 = 0 \qquad (5.3\text{--}19)$$

Now assume that the quintic equation 5.3–1 is irreducible in the field K. When the sextic 5.3–19 has no rational root, the Galois group of the quintic is A_5 or S_5 according to whether the discriminant $\Delta = 5^5(J^2 - 2^7 K)$ is a rational square or not. The quintic 5.3–1 is soluble by radicals in K if and only if $\Phi(\phi) = 0$ (equation 5.3–19) has a rational root so that the Galois group of the quintic is M_5, D_5, or C_5 (Table 2–3).

In the case where equation 5.3–19 has a rational root ϕ, the following procedure discussed by Berwick[4] can be used to determine whether the Galois group of the quintic is M_5, D_5, or C_5. Thus consider the function

$$\chi = (x_1 - x_2)^2(x_2 - x_3)^2(x_3 - x_4)^2(x_4 - x_5)^2(x_5 - x_1)^2 \qquad (5.3\text{--}20)$$

the factors of which correspond to the edges of the regular pentagon in Figure 5-1. The group of this function is the dihedral group D_5 whereas the group of the function

$$\sqrt{\chi} = (x_1 - x_2)(x_2 - x_3)(x_3 - x_4)(x_4 - x_5)(x_5 - x_1) \qquad (5.3\text{--}21)$$

is the cyclic group C_5. Now let ϕ be a rational root of the sextic 5.3–19 and χ a root of the following quadratic, which is rational because ϕ is rational:

$$\chi^2 - \phi\chi + 5^7(J^2 - 2^7 K) = 0 \qquad (5.3\text{--}22).$$

When this quadratic (or each of them if there are two) has no rational root, the Galois group of the corresponding quintic is the metacyclic group M_5. However, when the quadratic 5.3–22 has a rational root χ, the Galois group of the corresponding quintic is the dihedral group D_5 unless one of the roots of the quadratic is a rational square in which case the Galois group of the quintic is the cyclic group C_5.

Chapter 6

The Kiepert Algorithm for Roots of the General Quintic Equation

6.1 Introduction

The previous section (Section 5.3) discusses the problem of solving special quintic equations that are soluble by radicals, namely irreducible quintic equations with the soluble Galois groups C_5, D_5, and M_5. This chapter discusses the solution of general quintic equations including quintic equations having the Galois groups A_5 and S_5, which are not soluble by radicals.

The least complicated candidate for a quintic equation not solvable by radicals is the one-parameter equation

$$x^5 + ax + a = 0 \tag{6.1-1}$$

The roots of (6.1–1) can be expressed[1] as the Weierstrass elliptic function $\wp(z_j | 0, a)$, $1 \leq j \leq 5$, (Section 4.2), in which

$$[\wp'(z | g_2, g_3)]^2 = 4[\wp(z | g_2, g_3)]^3 - g_2 \wp(z | g_2, g_3) - g_3 \tag{6.1-2}$$

and z_j ($1 \leq j \leq 5$) are the five solutions of

$$\wp(z | 0, a) \wp'(z | 0, a) + i\sqrt{a}\, \wp(z | 0, a) + 2i\sqrt{a} = 0 \tag{6.1-3}$$

in the period parallelogram of the elliptic function.

These apparently elementary ideas are far from sufficient for a workable algorithm for solution of the general quintic equation

$$x^5 + Ax^4 + Bx^3 + Cx^2 + Dx + E = 0 \tag{6.1-4}$$

because of the following difficulties:

(1) Equation 6.1–1 is obtained by the trivial Tschirnhausen transformation (Section 3.3) $\xi = (\beta/\alpha)x$ from the Bring-Jerrard normal form of the quintic equation, namely

$$\xi^5 + \alpha\xi + \beta = 0 \tag{6.1-5}$$

so that $a = \dfrac{\alpha^5}{\beta^4} \neq 0$ \hspace{2em} (6.1–6).

---

[1] A. Hausner, The Bring-Jerrard Equation and Weierstrass Elliptic Functions, Amer. Math. Monthly, **69**, 193-196 (1962).

However, transformation of the original general quintic equation (6.1-4) to the Bring-Jerrard normal form (6.1-5) requires an additional Tschirnhausen transormation using two square roots and one root of a cubic equation (i.e., a total of three square roots and one cube root).[2] This Tschirnhausen transformation is very complicated.[2]

(2) Equation 6.1-3 must be solved in order to obtain the five values of z_j (1 ≤ j ≤ 5) in the period parallelogram in order to evaluate $\wp(z_j|0,a)$.

The difficult Tschirnhausen transformation from the general quintic (6.1-4) to its Bring-Jerrard normal form (6.1-5)[2] is avoided in the method described by Kiepert in 1878 for solution of the general quintic equation.[3] The author in collaboration with Prof. E. R. Canfield has verified the Kiepert algorithm by programming it onto a microcomputer and testing the program by determining the roots of some simple irreducible general quintic equations.[4,5]

The general structure of the Kiepert algorithm is depicted in Figure 6-1. The Kiepert algorithm may be broken down into the following seven steps:

(1) The Tschirnhausen transformation of the general quintic into the *principal quintic*

$$z^5 + 5az^2 + 5bz + c = 0 \qquad (6.1-7).$$

This Tschirnhausen transformation requires only a single square root in contrast to the Tschirnhausen transformation of the general quintic (6.1-4) to its Bring-Jerrard normal form (6.1-5) which requires three square roots and one cube root.

---

[2] A. Cayley, On Tschirnhausen's Transformation, *Phil. Trans. Roy. Soc. London*, **151**, 561-578 (1861); Cayley's collected works, paper 275.

[3] L. Kiepert, Auflösung der Gleichungen fünften Grades, *J. für Math.*, **87**, 114-133 (1878).

[4] R. B. King and E. R. Canfield, An Algorithm for Calculating the Roots of a General Quintic Equation from its Coefficients, *J. Math. Phys.*, **32**, 823-825 (1991).

[5] R. B. King and E. R. Canfield, Icosahedral Symmetry and the Quintic Equation, *Comput. and Math. with Appl.*, **24**, 13-28 (1992).

The Kiepert Algorithm for the Quintic Equation

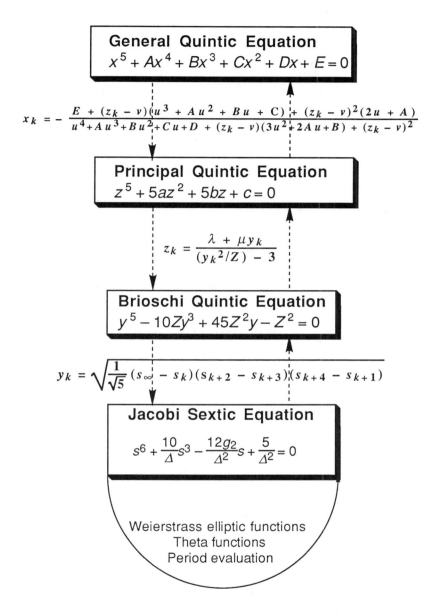

Figure 6–1: The general outline of the Kiepert algorithm for solution of the general quintic equation.

(2)   A second Tschirnhausen transformation of the principal quintic (6.1–7) into the *Brioschi quintic*

$$y^5 - 10Zy^3 + 45Z^2y - Z^2 = 0 \qquad (6.1-8)$$

in which the coefficients are expressed in terms of a single parameter Z. This Tschirnhausen transformation also requires only a single square root and is greatly facilitated by the use of polyhedral functions[6] and their special properties although Kiepert apparently did not recognize this point. Use of polyhedral functions leads to a modification of the Kiepert algorithm at one point in this step.

(3)   Transformation of the Brioschi quintic (6.1–8) into the *Jacobi sextic*

$$s^6 + \frac{10}{\Delta}s^3 - \frac{12g_2}{\Delta^2}s + \frac{5}{\Delta^2} = 0 \qquad (6.1-9)$$

in which

$$Z = -1/\Delta \qquad (6.1-10).$$

The roots, s_∞ and s_k ($0 \leq k \leq 4$), of the Jacobi sextic (6.1–9) can be used to calculate the roots, y_k, of the Brioschi quintic (6.1–8) by the relationship

$$y_k = \sqrt{\frac{1}{\sqrt{5}} (s_\infty - s_k)(s_{k+2} - s_{k+3})(s_{k+4} - s_{k+1})} \qquad (6.1-11)$$

with addition of indices modulo 5. The Galois group of the Jacobi sextic (6.1–9) is of order 60 like the A_5 Galois group of the quintic equations (6.1–4), (6.1–7), and (6.1–8) but involves transitive permutations of the six roots s_∞ and s_k ($0 \leq k \leq 4$) rather than only the five roots of the quintic equations.[7]

(4)   Solution of the Jacobi sextic (6.1–9) by the Weierstrass elliptic functions defined by the differential equation 6.1–2, followed by evaluation of the Weierstrass elliptic functions using genus 1 theta functions.

(5)   Evaluation of the periods of the theta functions corresponding to a particular Jacobi sextic (6.1–9). This requires solution of a cubic equation with coefficients simply calculable from the coefficients g_2 and Δ in (6.1–9). A simple function of the three roots of the cubic is then substituted into a special

---

[6]F. Klein, Vorlesungen über das Ikosaeder, Teubner, Leipzig, 1884.
[7]O. Perron, *Algebra*, Third Edition, de Gruyter, Berlin, 1951, Volume 2, Chapter 5.

infinite series called the Jacobi nome (Section 4.3).[8] Klein[9] has described an alternative approach to period evaluation using hypergeometric series.[10] Substitution of these periods into appropriate theta series then gives the roots s_∞ and s_k ($0 \leq k \leq 4$) of the Jacobi sextic (6.1–9).

(6) Calculation of the five roots, y_k, of the Brioschi quintic (6.1–8) from (6.1–11) followed by undoing the Tschirnhausen transformations of the quintic equations in the following sequence:

Brioschi (6.1–8) $\to$ Principal (6.1–7) $\to$ General (6.1–4).

Further details of these steps are given in the remainder of this chapter.

6.2 The Tschirnhausen Transformation of the General Quintic to the Principal Quintic

In order to transform the general quintic equation

$$x^5 + Ax^4 + Bx^3 + Cx^2 + Dx + E = 0 \tag{6.2–1}$$

into the principal quintic

$$z^5 + 5az^2 + 5bz + c = 0 \tag{6.2–2}$$

the following relationships derived from Newton's identities for the general quintic (6.1–4) are used:

$$\sum x_k = -A \tag{6.2–3a}$$

$$\sum x_k^2 = A^2 - 2B \tag{6.2–3b}$$

$$\sum x_k^3 = -A^3 + 3AB - 3C \tag{6.2–3c}$$

$$\sum x_k^4 = A^4 - 4A^2B + 4AC + 2B^2 - 4D \tag{6.2–3d}$$

For the Tschirnhausen transformation of (6.2–1) to (6.2–2) use

$$z = x^2 - ux + v \tag{6.2–4}$$

---

[8] A. Fletcher, *Guide to Tables of Elliptic Functions, Mathematical Tables and Other Aids to Computation*, **3**, 234 (1948).

[9] F. Klein, Über die Transformation der elliptischen funktionen und die Auflösung der Gleichungen fünften Grades, *Math. Ann.*, **14**, 111–172 (1878), *Gesammelte Mathematische Abhandlungen*, Springer-Verlag, Berlin, 1923, Volume 3, pp. 13-75.

[10] G. Sansone and J. Gerretsen, *Lectures on the Theory of Functions of a Complex Variable*, Woolters-Noordhoff, Groningen, 1969, Volume 2, Chapter 16.

and solve for u and v in this expression. From the absence of a z^4 term in the principal quintic (6.2–2) and equation 6.2–3a applied to (6.2–2) we get
$$0 = \Sigma z_k = \Sigma x_k^2 - u\Sigma x_k + 5v \qquad (6.2\text{–}5).$$
Substituting (6.2–3a) and (6.2–3b) into (6.2–5) leads to
$$5v = -Au - A^2 + 2B \qquad (6.2\text{–}6).$$
The absence of a z^3 term in the principal quintic (6.2–2) and equation (6.2–3b) applied to (6.2–2) leads to
$$0 = \Sigma z_k^2 = \Sigma(x_k^2 - ux_k + v)^2 \qquad (6.2\text{–}7).$$
Expanding the right hand side of equation 6.2–7 gives
$$0 = \Sigma x_k^4 + u^2 \Sigma x_k^2 + 5v^2 - 2u\Sigma x_k^3 + 2v\Sigma x_k^2 - 2uv\Sigma x_k \qquad (6.2\text{–}8).$$
Substituting (6.2–3a–d) and (6.2–6) into (6.2–8) gives the following quadratic equation in u:
$$(2A^2 - 5B)u^2 + (4A^3 - 13AB + 15C)u + (2A^4 - 8A^2B + 10AC + 3B^2 - 10D) = 0$$
$$(6.2\text{–}9)$$
Solving (6.2–9) for u and substituting u and v into the following expressions gives the coefficients a, b, and c of the principal quintic (6.2–2):
$$5a = -C(u^3 + Au^2 + Bu + C) + D(4u^2 + 3Au + 2B) - E(5u + 2A) - 10v^3 \qquad (6.2\text{–}10a)$$
$$5b = D(u^4 + 4u^3 + Bu^2 + Cu + D) - E(5u^3 + 4Au^2 + 3Bu + C) - 5v^4 - 10av \qquad (6.2\text{–}10b)$$
$$c = -E(u^5 + Au^4 + Bu^3 + Cu^2 + Du + E) - v^5 - 5av^2 - 5bv \qquad (6.2\text{–}10c)$$
In order to derive these equations define A', B', C', D', and E' as coefficients of the u-modified quintic equation whose roots are the roots of (6.2–1) minus u:
$$x^5 + A'x^4 + B'x^3 + C'x^2 + D'x + E' =$$
$$(x + u)^5 + A(x + u)^4 + B(x + u)^3 + C(x + u)^2 + D(x + u) + E = 0$$
$$(6.2\text{–}11)$$
In (6.2–11) x is defined by the original general quintic equation (6.2–1) and u by the Tschirnhausen transformation (6.2–4) and (6.2–9). Then the equations for the coefficients a, b, and c of the principal quintic (6.2–2) can be defined in reverse order as follows:

c (**equation 6.2–10c**): Write (6.2–2) as a product of linear factors using v as the variable to give
$$v^5 + 5av^2 + 5bv + c = \prod(v - z_k) \qquad (6.2\text{–}12).$$
Substituting equation (6.2–4) into equation (6.2–12) gives
$$v^5 + 5av^2 + 5bv + c = -\prod(x_k^2 - ux_k) = -\prod x_k \prod(x_k - u) = -EE' \qquad (6.2\text{–}13).$$
However,

The Kiepert Algorithm for the Quintic Equation

$$E' = E + Du + Cu^2 + Bu^3 + Au^4 + u^5 \tag{6.2-14}.$$

Substituting (6.2–14) into (6.2–13) and solving for c gives (6.2–10c).

b (equation 6.2–10b): Differentiate (6.2–12) to give

$$5v^4 + 10av + 5b = \sum_j \prod_{k \neq j} (v - z_k) \tag{6.2-15}$$

$$= \sum_j \prod_{k \neq j} x_k(x_k - u) \quad \text{(from 6.2–4)} \tag{6.2-16}$$

$$= \prod_k x_k(x_k - u) \sum_j \frac{1}{x_j(x_j - u)} \tag{6.2-17}$$

$$= EE' \left[\sum_j \left(\frac{1}{x_j - u} - \frac{1}{x_j} \right) \right] \frac{1}{u} \tag{6.2-18}$$

But

$$\sum_j \frac{1}{x_j} = \frac{D}{-E} \tag{6.2-19}$$

and

$$\sum_j \frac{1}{x_j - u} = \frac{D'}{-E'} = \frac{5u^4 + 4Au^3 + 3Bu^2 + 2Cu + D}{-(u^5 + Au^4 + Bu^3 + Cu^2 + Du + E)} \tag{6.2-20}.$$

Substituting (6.2–19) and (6.2–20) into (6.2–18) followed by routine algebra gives (6.2–10b).

a (equation 6.2–10a): Differentiate (6.2–12) twice to give

$$10v^3 + 5a = -\sum_{j_1 < j_2} \prod_{k \neq j_1, j_2} x_k(x_k - u) \tag{6.2-21}$$

$$= -\prod_k x_k(x_k - u) \sum_{j_1 < j_2} \frac{1}{x_{j_1}(x_{j_1} - u) x_{j_2}(x_{j_2} - u)} \tag{6.2-22}.$$

Multiplying $\frac{1}{x_k(x_k-u)} = \frac{1}{u}\left(\frac{1}{x_k-u} - \frac{1}{x_k}\right)$ for $k = j_1$ and j_2 to obtain a partial fraction decomposition of the summand gives

$$10v^3 + 5a =$$

$$-\prod_k x_k(x_k - u) \left(\frac{1}{u^2}\right) \left[\sum_{j_1<j_2}\left(\frac{1}{x_{j1}x_{j2}} + \frac{1}{(x_{j1} - u)(x_{j2} - u)}\right) - \sum_{j_1 \neq j_2}\frac{1}{x_{j2}(x_{j2}-u)}\right]$$

(6.2-23)

The terms in (6.2–23) are evaluated in terms of the coefficients of (6.2–1) and (6.2–11) and using (6.2–19) and (6.2–20) as follows:

$$\sum_{j_1<j_2} \frac{1}{x_{j1}x_{j2}} = \frac{-C}{-E} = \frac{C}{E} \qquad (6.2–24)$$

$$\sum_{j_1<j_2} \frac{1}{(x_{j1} - u)(x_{j2} - u)} = \frac{-C'}{-E'} = \frac{C'}{E'} = \frac{10u^3+6Au^2+3Bu+C}{u^5+Au^4+Bu^3+Cu^2+Du+E} \qquad (6.2–25)$$

$$\sum_{j_1 \neq j_2} \frac{1}{x_{j1}(x_{j2} - u)} = \left(\frac{DD'}{EE'} - \sum_j \frac{1}{x_j(x_j - u)}\right) = \frac{DD'}{EE'} - \frac{1}{u}\left(\frac{D'}{-E'} - \frac{D}{-E}\right) \qquad (6.2–26)$$

Combining

$$10v^3 + 5a = -\frac{1}{u^2}EE'\left\{\frac{C}{E} + \frac{C'}{E'} - \left[\frac{DD'}{EE'} - \frac{1}{u}\left(\frac{D'}{-E'} - \frac{D}{-E}\right)\right]\right\} \quad (6.2–27)$$

$$= \frac{1}{u^2}\left(DD' + \frac{1}{u}(D'E - DE') - CE' - C'E\right) \qquad (6.2–28)$$

Now,

$$\frac{1}{u}(D'E - DE') = -Du^4 + (5E - AD)u^3 + (4AE - BD)u^2 + (3BE - CD)u + (2CE - D^2)$$

(6.2–29)

Substituting (6.2–29) into (6.2–28) and using the values for C', D', and E' given in (6.2–14), (6.2–20), and (6.2–25) gives (6.2–10a).

6.3. The Tschirnhausen Transformation of the Principal Quintic to the Brioschi Quintic

The objective of this step is the transformation of the principal quintic (6.2–2) to a Brioschi quintic of the form

$$y^5 - 10Zy^3 + 45Z^2y - Z^2 = 0 \qquad (6.3\text{–}1)$$

in which all of the coefficients can be expressed in terms of a single parameter Z. This Tschirnhausen transformation uses the following relationship between the variables in (6.2–2) and (6.3–1):

$$z_k = \frac{\lambda + \mu y_k}{(y_k^2/Z) - 3} \qquad (6.3\text{–}2)$$

This Tschirnhausen transformation is best described geometrically using polyhedral functions on a Riemann sphere (Section 2.5) and is closely related to the partition of an object of icosahedral symmetry into five equivalent objects of tetrahedral symmetry (Section 2.3 and Figures 2.5 and 2.6). Details have been presented by Dickson[11] in 1930 and in a recent paper by the author[5] on the relationship between the quintic equation and icosahedral symmetry.

Consider a regular octahedron and a regular icosahedron represented by points on the surface of such a Riemann sphere so that the north pole is one of the vertices in each case. The polyhedral polynomials for the regular octahedron and icosahedron in these orientations in homogeneous form are as follows (Section 2.5) where x is taken to be u/v:

A. Octahedron (O_h symmetry):

Vertices: $\tau = uv(u^4 - v^4)$ $\qquad (6.3\text{–}3a)$

Edges: $\chi = u^{12} - 33u^8v^4 - 33u^4v^8 + v^{12}$ $\qquad (6.3\text{–}3b)$

Faces: $W = u^8 + 14u^4v^4 + v^8$ $\qquad (6.3\text{–}3c)$

B. Icosahedron (I_h symmetry):

Vertices: $f = uv(u^{10} + 11u^5v^5 - v^{10})$ $\qquad (6.3\text{–}4a)$

Edges: $T = u^{30} + 522u^{25}v^5 - 10005u^{20}v^{10} - 10005u^{10}v^{20} - 522u^5v^{25} + v^{30}$ $\qquad (6.3\text{–}4b)$

Faces: $H = -u^{20} + 228u^{15}v^5 - 494u^{10}v^{10} - 228u^5v^{15} - v^{20}$ $\qquad (6.3\text{–}4c)$

---

[11] L. E. Dickson, *Modern Algebraic Theories*, Sanborn, Chicago, 1930, Chapter 13.

The special symmetries of the octahedron and icosahedron lead to the following identities (Section 2.5):

Octahedron: $108\tau^4 - W^3 + \chi^2 \equiv 0$ (degree 24)   (6.3–5a)

Icosahedron: $1728f^5 - H^3 - T^2 \equiv 0$ (degree 60)   (6.3–5b)

Equation 6.3–5b plays a role in the Tschirnhausen transformation for conversion of a principal quintic (6.2–2) to the corresponding Brioschi quintic (6.3–1).

Consider further the icosahedron on a Riemann sphere so that one vertex is located at the north pole. The 30 edges of the icosahedron can be divided into five sets of six edges each (Figure 2.5) so that the midpoints of the edges define five octahedra whose vertex functions t_k and face functions W_k are as follows where $\varepsilon = \exp(2\pi i/5)$ and $0 \leq k \leq 4$:

$$t_k = \varepsilon^{3k}u^6 + 2\varepsilon^{2k}u^5v - 5\varepsilon^k u^4v^2 - 5\varepsilon^{4k}u^2v^4 - 2\varepsilon^{3k}uv^5 + \varepsilon^{2k}v^6 \quad (6.3\text{–}6a)$$

$$W_k = -\varepsilon^{4k}u^8 + \varepsilon^{3k}u^7v - 7\varepsilon^{2k}u^6v^2 - 7\varepsilon^k u^5v^3 + 7\varepsilon^{4k}u^3v^5 - 7\varepsilon^{3k}u^2v^6 - \varepsilon^{2k}uv^7 - \varepsilon^k v^8 \quad (6.3\text{–}6b)$$

The vertex functions t_k are the five roots of a Brioschi quintic equation

$$t^5 - 10ft^3 + 45f^2 t - T = 0 \quad (6.3\text{–}7)$$

in which f and T are the vertex and edge functions of the icosahedron defined by (6.3–4a) and 6.3–4b), respectively. Equation 6.3–1 is a special case of equation 6.3–7 in which $f^2 = T$ so that $t = yT/f^2$.

Now consider the principal quintic (6.2–2). Express its roots as the polyhedral functions

$$z_k = \left(\frac{\lambda f}{H}\right) W_k + \left(\frac{\mu f^3}{HT}\right) t_k W_k \quad (6.3\text{–}8).$$

In (6.3–8) f, H, and T are the icosahedral functions (6.3–4) and t_k and W_k are functions (equations 6.3–6a and 6.3–6b) of the five octahedra formed by the midpoints of the 30 icosahedral edges (Figure 2.5). The right hand side of (6.3–8) is thus homogeneous of degree 0 in the variable pair u, v. The ability to express the roots of any *principal* quintic (6.2–2) by (6.3–8) depends upon the following properties of the polyhedral functions (6.3–6) which arise from the impossibility of any product of the icosahedral functions f, H, and T, of degrees 12, 20, and 30, respectively, having degrees 8, 14, 16, 22, or 28 and which lead to the vanishing of the z^4 and z^3 coefficients in the principal quintic (6.2–2):

The Kiepert Algorithm for the Quintic Equation

$$\Sigma W_k = \Sigma t_k W_k = \Sigma W_k^2 = \Sigma t_k W_k^2 = \Sigma t_k^2 W_k^2 = 0 \qquad (6.3\text{--}9).$$

The next step is the calculation of the coefficients a, b, and c of the principal quintic (6.2–2) with roots z_k expressed by (6.3–8) in terms of the two composite icosahedral functions Z and V where

$$Z = \frac{f^5}{T^2} \qquad (6.3\text{--}10)$$

$$V = \frac{H^3}{f^5} \qquad (6.3\text{--}11)$$

so that Z and V are related by

$$\frac{1}{Z} + V = 1728 \qquad (6.3\text{--}12)$$

derived from the identity 6.3–5b. This calculation uses the powers t_k^n and W_k^n ($2 \leq n \leq 4$) calculated by brute force from (6.3–6a) and (6.3–6b) although only the first few terms of each power are required as well as the relationship

$$\sum_{k=1}^{5} \varepsilon^{nk} = 0 \qquad (6.3\text{--}13).$$

By comparing the indicated terms with the indicated products of the icosahedral functions f, H, and T (equation 6.3–4) the equations 6.3–14 and 6.3–15 can be derived as follows:

$u^{22}v^2$ terms: $\Sigma W_k^3 = (5)(-24)f^2 = -120f^2$ (6.3–14a)

u^{30} terms: $\Sigma W_k^3 t_k = (5)(-1)T = -5T$ (6.3–14b)

$u^{33}v^3$ terms: $\Sigma W_k^3 t_k^2 = (5)(20-18-96-22)u^{33}v^3 = (5)(-72)f^3 = -360f^3$ (6.3–14c)

$u^{41}v$ terms: $\Sigma W_k^3 t_k^3 = (5)(-6+3)u^{41}v = (5)(-3)fT = -15fT$ (6.3–14d)

$u^{31}v$ terms: $\Sigma W_k^4 = (5)(-4)u^{31}v = (5)(4)fH = 20\,fH$ (6.3–15a)

$\Sigma W_k^4 t_k = 0$ since no product of f (degree 12), H (degree 20), and T (degree 30) can have degree 38.

$u^{42}v^2$ terms: $\Sigma W_k^4 t_k^2 = (5)(34-16-6)u^{42}v^2 = (5)(-12)f^2H = -60f^2H$ (6.3–15b)

u^{50} terms: $\Sigma W_k^4 t_k^3 = (5)(-1)HT = -5HT$ (6.3–15c)

$u^{53}v^3$ terms: $\Sigma W_k^4 t_k^4 = (5)(-88-16+272-60)u^{53}v^3 = (5)(-108)f^3H = -540f^3H$

    (6.3–15d)

In addition

$\Sigma z_k^3 = -15a$ (6.3–16a)

$\Sigma z_k^4 = -20b$ (6.3–16b)

Calculating Σz_k^3 and Σz_k^4 by expanding (6.3–8) and substituting the values from (6.3–10), (6.3–14), (6.3–15), and (6.3–16) gives

$$Va = 8\lambda^3 + \lambda^2\mu + (72\lambda\mu^2 + \mu^3)Z \qquad (6.3-17)$$
$$Vb = -\lambda^4 + 18\lambda^2\mu^2 Z + \lambda\mu^3 Z + 27\mu^4 Z^2 \qquad (6.3-18)$$

Now consider (6.3–7) and its five roots t_k to give

$$x^5 - 10fx^3 + 45f^2 x - T = \prod(x - t_k) \qquad (6.3-19)$$

for all values of x. Now let x be $-\lambda T/\mu f^2$ which generates the Brioschi quintic (6.3–1) if $y = -\lambda/\mu$.

Then

$$\prod(\lambda T + \mu f^2 t_k) = \lambda^5 T^5 - 10\lambda^3\mu^2 T^3 f^5 + 45\lambda\mu^4 T f^{10} + \mu^5 f^{10} T \qquad (6.3-20).$$

However, $\prod W_k = -H^2 \qquad (6.3-21)$

so that substituting the values for V and Z from equations 6.3–10 and 6.3–11 gives

$$Vc = \lambda^5 - 10\lambda^3\mu^2 Z + 45\lambda\mu^4 Z^2 + \mu^5 Z^2 \qquad (6.3-22)$$

Thus the values of a, b, and c obtained from (6.3–17), (6.3–18), and (6.3–22) determine the coefficients of the principal quintic (6.2–2) corresponding to the Brioschi quintic (6.3–7) with the changes of variables (6.3–2).

The equations 6.3–17, 6.3–18, and 6.3–22 obtained above can now be inverted so that the parameters λ, μ, Z, and V can be calculated for the Brioschi quintic (6.3–1) corresponding to a given principal quintic (6.2–2) with coefficients a, b, and c. First add λVb from (6.3–18) to Vc from equation (6.3–22) to give $\mu^2 ZVa$ as determined by (6.3–17), i.e.,

$$\mu^2 Za = \lambda b + c \qquad (6.3-23).$$

Then subtract $\mu^2 ZVb$ (6.3–18) from λVc (6.3–22) and use (6.3–23) to give

$$V = \frac{(\lambda^2 - 3\mu^2 Z)^3}{\lambda c - \mu^2 Z b} = \frac{(a\lambda^2 - 3b\lambda - 3c)^3}{a^2(\lambda ac - \lambda b^2 - bc)} \qquad (6.3-24)$$

Now combine equations (6.3–17) and (6.3–18) in the indicated manner to give

$$V\left(\frac{\lambda a + 8b}{\mu}\right) = \lambda^3 + 216\lambda^2\mu Z + 9\lambda\mu^2 Z + 216\mu^3 Z^2 \qquad (6.3-25)$$

Divide the square of the right hand side of (6.3–25) by Z and subtract the result from 27 times the square of the right hand side of (6.3–17) and combine pairs of terms related by the identity (6.3–12) to give

$$27a^2 V - \frac{V(\lambda a + 8b)^2}{\mu^2 Z} = (\lambda^2 - 3\mu^2 Z)^3 \qquad (6.3-26).$$

Now substitute (6.3–24) into (6.3–26) to give

$$27a^2 - \frac{(\lambda a + 8b)^2}{\mu^2 Z} = \lambda c - \mu^2 Z b \qquad (6.3\text{--}27).$$

Substitution of $(\lambda b + c)/a$ for $\mu^2 Z$ from (6.3–23) now gives the following quadratic equation for λ in terms only of the coefficients a, b, and c of the principal quintic (6.2–2):

$$\lambda^2(a^4 + abc - b^3) - \lambda(11a^3 b - ac^2 + 2b^2 c) + 64a^2 b^2 - 27a^3 c - bc^2 = 0 \qquad (6.3\text{--}28)$$

After solving this quadratic for λ, the value for V can be found by (6.3–24). After obtaining λ and V, equation 6.3–17 can be rewritten as

$$(\lambda^2 + \mu^2 Z)\mu = Va - 8\lambda^3 - 72\lambda\mu^2 Z \qquad (6.3\text{--}29)$$

Substitution of (6.3–23) for $\mu^2 Z$ and solving for μ then gives

$$\mu = \frac{Va^2 - 8\lambda^3 a - 72\lambda^2 b - 72\lambda c}{\lambda^2 a + \lambda b + c} \qquad (6.3\text{--}30)$$

In this way the four parameters λ, μ, V, and Z are obtained for a Brioschi quintic (6.3–1) with roots (6.3–8) corresponding to a principal quintic (6.2–2) with coefficients a, b, and c.

There remains the problem of deriving the Tschirnhausen transformation (6.3–2) relating the principal quintic (6.2–2) to the one-parameter Brioschi quintic (6.3–1). The functions W_k corresponding to the octahedron vertex functions t_k which are solutions of the more general Brioschi quintic (6.3–7) vanish at the midpoints of the faces of the octahedra which are located at the midpoints of the faces of the underlying icosahedron where H also vanishes. Hence each W_k is a factor of H. In addition, transposing the term T of the quintic (6.3–7), squaring both sides, and replacing t^2 by $3f$ gives

$$1728 f^5 = T^2 \qquad (6.3\text{--}31)$$

Substituting the icosahedral identity (6.3–5b) into (6.3–31) gives $H = 0$, indicating that $t_k^2 - 3f$ is a factor of H for each k and so

$$H = W_k(t_k^2 - 3f) \qquad 0 \le k \le 4 \qquad (6.3\text{--}32).$$

Substituting (6.3–32) into (6.3–8), suppressing subscripts, using (6.3–10) for Z, and the relationship $t = yT/f^2$ to convert the two-parameter Brioschi quintic (6.3–7) to the one-parameter Brioschi quintic (6.3–1) gives the required relationship (6.3–2) between the variable z of the principal quintic and the variable y of the Brioschi quintic (6.3–1).

6.4. Transformation of the Brioschi Quintic to the Jacobi Sextic

The objective of this step is the transformation of the Brioschi quintic (6.3–1) into a Jacobi sextic

$$s^6 - 10fs^3 + Hs + 5f^2 = 0 \qquad (6.4\text{–}1)$$

whose roots can be expressed in terms of elliptic functions. The Galois group of (6.4–1) like that of the Brioschi quintic (6.3–1) or (6.3–7) is a transitive group of order 60 but now involves permutations on six objects (the roots of equation 6.4–1) rather than five objects (the roots of equation 6.3–1 or 6.3–12). If one of the roots of the Jacobi sextic is designated as s_∞, the five other roots s_k ($0 \le k \le 4$) have the special form

$$s_k = \left(\frac{s_\infty}{5}\right)\left(1 + r\varepsilon^k - \frac{1}{r\varepsilon^k}\right)^2 \qquad (6.4\text{–}2)$$

where ε is any primitive fifth root of unity. From here on we shall take ε to be $\exp(2\pi i/5)$.

In order to effect this transformation the following theorem of Perron[7] is used:

Theorem 6.4–1 (Perron): In order to find the roots of the Brioschi quintic

$$t^5 - 10ft^3 + 45f^2 t - T = 0 \qquad (6.4\text{–}3)$$

the quantity H for the corresponding Jacobi sextic equation (6.4–1) is determined to satisfy the icosahedral identity

$$1728f^5 - H^3 - T^2 = 0 \qquad (6.4\text{–}4).$$

If the roots of this Jacobi sextic are s_∞ and s_k ($0 \le k \le 4$), then the roots t_k of the Brioschi quintic (6.4–3) are

$$t_k = \sqrt{\frac{1}{\sqrt{5}}(s_\infty - s_k)(s_{k+2} - s_{k+3})(s_{k+4} - s_{k+1})} \qquad (6.4\text{–}5).$$

The $\sqrt{5}$ in (6.4–5) arises from the relationship

$$\varepsilon^4 + \varepsilon - \varepsilon^2 - \varepsilon^3 = \sqrt{5} \qquad (6.4\text{–}6)$$

and the sign of the outer square roots can be obtained by writing the Brioschi quintic as

$$t = \frac{T}{t^4 - 10ft^2 + 45f^2} \qquad (6.4\text{–}7)$$

The Kiepert Algorithm for the Quintic Equation

so that only even powers of t appear on the right hand side. Choice of a different primitive fifth root of unity for ε may lead to replacement of $\sqrt{5}$ on the right side of (6.4–6) by $-\sqrt{5}$.

Proof: Consider a special quintic equation

$$\prod_{k=0}^{4}(\eta - \eta_k) = \eta^5 + a_1\eta^4 + a_2\eta^3 + a_3\eta^2 + a_4\eta + a_5 = 0 \quad (6.4\text{–}8)$$

in which the roots have the form

$$c_1\varepsilon^k + c_2\varepsilon^{2k} + c_3\varepsilon^{3k} + c_4\varepsilon^{4k} = \eta_k \quad (0 \leq k \leq 4) \quad (6.4\text{–}9).$$

From Newton's identities (e.g., equations 6.2–3) the coefficients a_m ($1 \leq m \leq 5$) are found to be

$$a_1 = 0 \quad (6.4\text{–}10\text{a})$$

$$a_2 = -5(c_1c_4 + c_2c_3) \quad (6.4\text{–}10\text{b})$$

$$a_3 = -5(c_1^2c_3 + c_2^2c_1 + c_3^2c_4 + c_4^2c_2) \quad (6.4\text{–}10\text{c})$$

$$a_4 = -5(c_1^3c_2 + c_2^3c_4 + c_3^3c_1 + c_4^3c_3) + 5(c_1^2c_4^2 + c_2^2c_3^2) - 5c_1c_2c_3c_4 \quad (6.4\text{–}10\text{d})$$

$$a_5 = -(c_1^5 + c_2^5 + c_3^5 + c_4^5) + 5(c_1^3c_3c_4 + c_2^3c_1c_3 + c_3^3c_2c_4 + c_4^3c_1c_2) -$$
$$5(c_1^2c_2^2c_4 + c_1^2c_3^2c_2 + c_2^2c_4^2c_3 + c_3^2c_4^2c_1) \quad (6.4\text{–}10\text{e})$$

Application of equations (6.4–10) in the special case

$$c_1 = r; \ c_2 = 0; \ c_3 = 0; \ c_4 = -1/r \quad (6.4\text{–}11)$$

leads to

$$\prod_{k=0}^{4}\left[\eta - \left(r\varepsilon^k - \frac{1}{r\varepsilon^k}\right)\right] = \eta^5 + 5\eta^3 + 5\eta - \left(r^5 - \frac{1}{r^5}\right) \quad (6.4\text{–}12)$$

or if $\eta - 1$ is used instead of η

$$\prod_{k=0}^{4}\left[\eta - \left(1 + r\varepsilon^k - \frac{1}{r\varepsilon^k}\right)\right] = \eta^5 - 5\eta^4 + 15\eta^3 - 25\eta^2 + 25\eta - 125\sigma \quad (6.4\text{–}13)$$

where $125\sigma = 11 + r^5 - \dfrac{1}{r^5} \quad (6.4\text{–}14)$.

Replacing η by $-\eta$, multiplying the resulting equation by (6.4–13), and changing the sign gives

$$\prod_{k=0}^{4}\left[\eta + \left(1 + r\varepsilon^k - \frac{1}{r\varepsilon^k}\right)\right]\left[\eta - \left(1 + r\varepsilon^k - \frac{1}{r\varepsilon^k}\right)\right]$$

$$= \prod_{k=0}^{4}\left[\eta^2 - \left(1 + r\varepsilon^k - \frac{1}{r\varepsilon^k}\right)^2\right]$$

$$= \eta^{10} + 5\eta^8 + 25\eta^6 + 125(1 - 10\sigma)\eta^4 + 625(1 - 10\sigma)\eta^2 - 125^2\sigma^2 \quad (6.4\text{–}15)$$

Now replace η^2 by $5s/\rho$ and multiply first by $(\rho/5)^5$ and then by $s - \rho$ to give

$$(s - \rho)\prod_{k=0}^{4}\left[s - \frac{\rho}{5}\left(1 + r\varepsilon^k - \frac{1}{r\varepsilon^k}\right)^2\right]$$

$$= s^6 - 10\sigma\rho^3 s^3 - (1 - 10\sigma + 5\sigma^2)\rho^5 s + 5\sigma^2\rho^6 \quad (6.4\text{–}16)$$

Equation 6.4–16 set to zero corresponds to the Jacobi sextic (6.4–1) in which

$$f = \sigma\rho^3 \quad (6.4\text{–}17a)$$
$$H = -(1 - 10\sigma + 5\sigma^2)\rho^5 \quad (6.4\text{–}17b)$$

and the roots are

$$s_\infty = \rho \text{ and } s_m = \frac{\rho}{5}\left(1 + r\varepsilon^m - \frac{1}{r\varepsilon^m}\right)^2 \quad (0 \le m \le 4) \quad (6.4\text{–}18)$$

Now consider the following five quantities which will be used to derive the five roots of a quintic in v ($0 \le k \le 4$):

$$v_k = (s_\infty - s_k)(s_{k+2} - s_{k+3})(s_{k+4} - s_{k+1}) \quad (6.4\text{–}19).$$

Substituting (6.4–18) into (6.4–19) and using (6.4–6) and (6.4–17a) where appropriate gives

$$\frac{v_k}{\sqrt{5}} - 4f =$$

$$\frac{\rho^3}{125}\left[\left(r^6 - 24r - \frac{6}{r^4}\right)\varepsilon^k + \left(15r^2 + \frac{20}{r^3}\right)\varepsilon^{2k} + \left(-20r^3 + \frac{15}{r^2}\right)\varepsilon^{3k} + \left(-6r^4 + \frac{24}{r} + \frac{1}{r^6}\right)\varepsilon^{4k}\right]$$

$$(6.4\text{–}20)$$

Now consider the quintic polynomial

$$\prod_{k=0}^{4}\left[v - \left(\frac{v_k}{\sqrt{5}} - 4f\right)\right] = v^5 + a_1 v^4 + a_2 v^3 + a_3 v^2 + a_4 v + a_5 \quad (6.4\text{–}21)$$

The coefficients a_m of the polynomial have the form

The Kiepert Algorithm for the Quintic Equation   111

$$a_m = \sum_i \sum_j C_{ij} \rho^{3i+5j} \sigma^i (1 - 10\sigma + 5\sigma^2)^j \qquad (6.4\text{--}22)$$

where $3m = 3i + 5j$.

Considering (6.4–20) as a special case of (6.4–9) and applying the formulas 6.4–10 gives

$$\prod_{k=0}^{4}\left[v - \left(\frac{v_k}{\sqrt{5}} - 4f\right)\right] = v^5 + 30f^2v^3 + 100f^3v^2 + 105f^4v + 36f^5 + H^3 \qquad (6.4\text{--}23)$$

Now replace v by w where $v = \dfrac{w}{\sqrt{5}} - 4f \qquad (6.4\text{--}24)$

and multiply by $25\sqrt{5} = (\sqrt{5})^5$ to give

$$\prod_{k=0}^{4}(w - v_k) = w^5 - 20\sqrt{5}fw^4 + 950f^2w^3 - 4500\sqrt{5}f^3w^2 + 50625f^4w + 25\sqrt{5}(H^3 -$$

$$1728f^5) = 0 \qquad (6.4\text{--}25)$$

Now replace w by t where $w = \sqrt{5}t^2 \qquad (6.4\text{--}26)$

and divide by $25\sqrt{5}$ to give

$$t^{10} - 20ft^8 + 190f^2t^6 - 900f^3t^4 + 2025f^4t^2 + (H^3 - 1728f^5) = 0 \qquad (6.4\text{--}27)$$

which can be written as

$$(t^5 - 10ft^3 + 45f^2t)^2 = -H^3 + 1728f^5 \qquad (6.4\text{--}28)$$

Substitution of the icosahedral identity (6.4–4) in (6.4–28) and then taking the square root of each side gives the Brioschi quintic (6.4–3). Thus

$$t_k = \sqrt{\frac{v_k}{\sqrt{5}}} \qquad (6.4\text{--}29)$$

after considering the effects of the variable changes $v \to w \to t$ with the outer square root arising from equation (6.4–26).

This concludes the summary of Perron's proof. ∎

The next step is to relate the coefficients f, H, and T of the Brioschi quintic and the associated Jacobi sextic to the invariants g_2, g_3, and Δ of corresponding elliptic functions satisfying the following identity (compare equation 4.2–19g):

$$\Delta = g_2{}^3 - 27g_3{}^2 \tag{6.4-30}$$

Write the Brioschi quintic (6.3–7) as

$$t^5 + \frac{10}{\Delta}t^3 + \frac{45}{\Delta^2}t - \frac{216g_3}{\Delta^3} = 0 \tag{6.4-31}$$

so that $f = \dfrac{-1}{\Delta}$ \hfill (6.4–32)

and $T = \dfrac{216g_3}{\Delta^3}$ \hfill (6.4–33).

Applying the icosahedral identity (6.4–4) and the elliptic identity (6.4–30) leads to the relationship

$$\Delta = -\left(\frac{H\Delta^2}{12}\right)^3 - 27g_3{}^2 \tag{6.4-34}$$

and the Jacobi sextic corresponding to the Brioschi quintic (6.4–31) is

$$s^6 + \frac{10}{\Delta}s^3 - \frac{12g_2}{\Delta^2}s + \frac{5}{\Delta^2} = 0 \tag{6.4-35}.$$

Now let us relate the parameter Z in the one-parameter Brioschi quintic (6.3–1) to the elliptic invariants g_2 and Δ. Comparing (6.3–1) and (6.4–31) indicates that

$$Z = -\frac{1}{\Delta} \tag{6.4-36}$$

$$Z^2 = \frac{1}{\Delta^2} = \frac{216g_3}{\Delta^3} \Rightarrow g_3 = \frac{\Delta}{216} \tag{6.4-37}.$$

Substituting (6.4–37) in the elliptic identity (6.4–30) leads to

$$\frac{g_2{}^3}{\Delta} = \frac{\Delta}{1728} + 1 = \frac{-1}{1728Z} + 1 \tag{6.4-38}.$$

Defining V by the identity (6.3–11) then gives

$$V = \frac{1728 g_2{}^3}{\Delta} \tag{6.4-39}.$$

Thus the elliptic invariants Δ and g_2 can be obtained from the V and Z calculated in the previous section by the following simple relationships:

The Kiepert Algorithm for the Quintic Equation

$$\Delta = \frac{-1}{Z} \tag{6.4-40}$$

$$g_2 = \frac{\sqrt[3]{V\Delta}}{12} = \frac{1}{12}\sqrt[3]{\frac{1-1728Z}{Z^2}} \tag{6.4-41}$$

6.5 Solution of the Jacobi Sextic with Weierstrass Elliptic Functions

The solution of cubic equations (Section 5.1) using the (periodic) trigonometric functions makes use of their period division by three. For example, the solution of the cubic equation

$$x^3 + 3Hx + G = 0 \tag{6.5-1}$$

can be expressed as

$$x_k = 2\sqrt{-H}\,\cos\!\left(\frac{\phi + 2k\pi}{3}\right) \qquad (0 \le k \le 2) \tag{6.5-2}$$

where $\cos\phi = \dfrac{-G}{2\sqrt{-H^3}}$ \hfill (6.5-3).

Analogously the solution of the Jacobi sextic (6.4-1) corresponding to a Brioschi quintic (6.3-7) relates to the period division of the doubly periodic elliptic functions by five. Thus using a formula derived by Kiepert[12,13] the following equation will be solved (compare equation 4.2-28):

$$\frac{\sigma(5u)}{\sigma^{25}(u)} = \frac{1}{(2!3!4!)^2}\begin{vmatrix} \wp' & \wp'' & \wp''' & \wp^{IV} \\ \wp'' & \wp''' & \wp^{IV} & \wp^{V} \\ \wp''' & \wp^{IV} & \wp^{V} & \wp^{VI} \\ \wp^{IV} & \wp^{V} & \wp^{VI} & \wp^{VII} \end{vmatrix} = 0 \tag{6.5-4}$$

---

[12] K. Weierstrass and H. A. Schwarz, *Formeln und Lehrsätze zum Gebrauche der Elliptischen Funktionen*, Springer Verlag, Berlin, 1893, p 19.

[13] L. Kiepert, Wirkliche Ausführung der Ganzzahligen Multiplikation der Elliptischen Funktionen, *J. für Math.*, **76**, 21-33 (1873).

In (6.5-4) the entries in the 4×4 determinant are the first through seventh derivatives of the Weierstrass $\wp$ function (Section 4.2) defined by the differential equation

$$(\wp')^2 = 4\wp^3 - g_2\wp - g_3 \qquad (6.5\text{-}5)$$

and the Weierstrass σ function is defined by

$$\wp(u) = \frac{-d^2[\ln \sigma(u)]}{du^2} \quad \text{(compare equation 4.2-11c)} \qquad (6.5\text{-}6).$$

Since the zeroes of $\sigma(u)$ occur at $u + 2m\omega + 2n\omega'$, the determinant in (6.5-4) vanishes when

$$u_{mn} = \frac{2m\omega + 2n\omega'}{5} \qquad (6.5\text{-}7).$$

In equation (6.5-7) 2ω and $2\omega'$ are the two periods of the elliptic functions and m and n are integers mod 5 both of which are *not* zero. The period division by five in (6.5-7) is analogous to the period division by three of the cosine in (6.5-2) used to solve the cubic equation (6.5-1).

Evaluation of the determinant in (6.5-4) is found to provide a method for expressing the roots of the Jacobi sextic (6.4-1) in terms of the Weierstrass elliptic $\wp$ function (6.5-5). The periods of the elliptic functions can be calculated from the coefficients of the Jacobi sextic (6.4-1) or the corresponding Brioschi quintic (6.3-1) by either of two methods, one of which will be presented later in more detail.

In order to evaluate the determinant in (6.5-4), Perron[7] uses the following derivatives of $\wp(u)$ in addition to $\wp'(u)$ (equation 6.5-5):

$$\wp'' = 6\wp^2 - \frac{g_2}{2} \qquad (6.5\text{-}8b)$$

$$\wp''' = 12\wp\,\wp' \qquad (6.5\text{-}8c)$$

$$\wp^{IV} = 12(\wp')^2 + 12\wp\,\wp'' \qquad (6.5\text{-}8d)$$

$$\wp^{V} = 36\wp'\,\wp'' + 12\wp\,\wp''' \qquad (6.5\text{-}8e)$$

$$\wp^{VI} = 36(\wp'')^2 + 48\wp'\,\wp''' + 12\wp\,\wp^{IV} \qquad (6.5\text{-}8f)$$

$$\wp^{VII} = 120\wp''\,\wp''' + 60\wp'\,\wp^{IV} + 12\wp\,\wp^{V} \qquad (6.5\text{-}8g)$$

This sequence of derivatives of $\wp$ resembles Painlevé chains discussed elsewhere by the author.[14,15]

Several steps will now be taken in order to simplify the determinant in (6.5–4) before its explicit evaluation. First subtract $12\wp$ times the first column from the third column and subtract $12\wp$ times the second column and $12\wp'$ times the first column from the last (fourth) column after substituting the formulas for the derivatives (6.5–8c to 6.5–8g) to give

$$\begin{vmatrix} \wp' & \wp'' & 0 & 0 \\ \wp'' & \wp''' & 12(\wp')^2 & 24\wp'\wp'' \\ \wp''' & \wp^{IV} & 36\wp'\wp'' & 36(\wp'')^2+36\wp'\wp''' \\ \wp^{IV} & \wp^{V} & 36(\wp'')^2+48\wp'\wp''' & 120\wp'\wp'''+48\wp'\wp^{IV} \end{vmatrix} = 0 \qquad (6.5\text{–}9).$$

Now perform the same operation with the rows and suppress factors of 12 in the third and fourth columns to give

$$\begin{vmatrix} \wp' & \wp'' & 0 & 0 \\ \wp'' & \wp''' & (\wp')^2 & 2\wp'\wp'' \\ 0 & 12(\wp')^2 & 3\wp'\wp'' & 3(\wp'')^2+3\wp'\wp''' \\ 0 & 24\wp'\wp'' & 3(\wp'')^2+4\wp'\wp'''-12\wp(\wp')^2 & 10\wp''\wp'''+4\wp'\wp^{IV}-24\wp\wp'\wp'' \end{vmatrix} = 0$$

(6.5–10)

Now substitute the expressions (6.5–8c) and (6.5–8d) for $\wp'''$ and $\wp^{IV}$, respectively, and suppress factors of three in the third and fourth rows to give

$$\begin{vmatrix} \wp' & \wp'' & 0 & 0 \\ \wp'' & 12\wp\wp' & (\wp')^2 & 2\wp'\wp'' \\ 0 & 4(\wp')^2 & \wp'\wp'' & (\wp'')^2+12\wp(\wp')^2 \\ 0 & 8\wp'\wp'' & (\wp'')^2+12\wp(\wp')^2 & 48\wp\wp'\wp''+16(\wp')^3 \end{vmatrix} = 0 \qquad (6.5\text{–}11).$$

Multiplying out this determinant gives the following polynomial which is of degree 12 in $\wp$ after considering (6.5–5) and (6.5–8b):

---

[14] R. B. King, Systematics of Strongly Self-Dominant Higher Order Differential Equations Based on the Painlevé Analysis of their Singularities, *J. Math. Phys.*, **27**, 966–971 (1986).

[15] R. B. King, Painlevé Chains for the Study of Integrable Higher-Order Differential Equations, *J. Phys. A.: Math. Gen.*, **20**, 2333–2345 (1987).

$$[(\wp")^2 - 12\wp(\wp')^2]^3 - 16(\wp')^4 \wp"[(\wp")^2 - 12\wp(\wp')^2] - 64(\wp')^8 = 0 \qquad (6.5\text{--}12).$$

The 24 values of u_{mn} in (6.5–7) relate to the 12 roots $\wp_{m,n}$ of (6.5–12) in pairs by the relationship

$$\wp_{-m,-n} = \wp_{m,n} \qquad (6.5\text{--}13)$$

using mod 5 arithmetic on the subscripts.

Next Perron evaluates

$$y_{mn} = \wp_{2m,2n} - \wp_{m,n} = \wp\left(\frac{4m\omega + 4n\omega}{5}\right) - \wp\left(\frac{2m\omega + 2n\omega}{5}\right) \qquad (6.5\text{--}14)$$

and uses the following formula from Weierstrass:

$$\wp(u) - \wp(2u) = \frac{1}{4}\frac{d^2}{du^2}\log \wp'(u) = \frac{\wp'(u)\wp'''(u) - \wp''(u)^2}{4\wp'(u)^2} \qquad (6.5\text{--}15).$$

Substituting equations (6.5–8c) and (6.5–14) into (6.5–15) gives

$$y_{mn} = \frac{(\wp"_{m,n})^2 - 12\wp_{m,n}(\wp'_{m,n})^2}{4(\wp'_{m,n})^2} \qquad (6.5\text{--}16)$$

for all integers m and n mod 5 excluding $m = n = 0$. Substituting (6.5–16) into (6.5–12) and dividing by $64(\wp')^6$ gives the cubic equation

$$y^3 - \wp"y - (\wp')^2 = 0 \qquad (6.5\text{--}17).$$

Also solving (6.5–16) for $(\wp")^2$ gives

$$(\wp")^2 = (4y + 12\wp)(\wp')^2 \qquad (6.5\text{--}18).$$

Combining (6.5–17) and (6.5–18) gives

$$(\wp")^2 = (4y + 12\wp)(y^3 - \wp"y) \qquad (6.5\text{--}19).$$

Substituting the expressions for $(\wp')^2$ (equation 6.5–5) and $\wp"$ (equation 6.5–8b) into (6.5–17) and (6.5–19) gives the pair of equations:

$$y^3 - (6\wp^2 - \tfrac{1}{2}g_2)y - (4\wp^3 - g_2\wp - g_3) = 0 \qquad (6.5\text{--}20a)$$

$$(6\wp^2 - \tfrac{1}{2}g_2)^2 = (4y + 12\wp)(y^3 - 6\wp^2 y + \tfrac{1}{2}g_2 y) \qquad (6.5\text{--}20b)$$

Perron[7] now proceeds further by introducing a new variable z where

$$\wp = \tfrac{1}{2}(z - y) \qquad (6.5\text{--}21).$$

The Kiepert Algorithm for the Quintic Equation

Substituting (6.5–21) into (6.5–20a) and (6.5–20b) and ordering according to powers of z gives

$$z^3 - (3y^2 + g_2)z - 2g_3 = 0 \qquad (6.5\text{–}22\text{a})$$
$$9z^4 - 6(5y^2 + g_2)z^2 + 5y^4 - 2g_2y^2 + g_2^2 = 0 \qquad (6.5\text{–}22\text{b}).$$

Solving (6.5–22b) for z^2 by using the quadratic formula gives

$$3z^2 = 5y^2 + g_2 + 2y\sqrt{5y^2 + 3g_2} \qquad (6.5\text{–}23).$$

Writing (6.5–22a) as

$$z(z^2 - 3y^2 - g_2) = 2g_3 \qquad (6.5\text{–}24)$$

and squaring gives

$$z^2(z^2 - 3y^2 - g_2)^2 = 4g_3^2 \qquad (6.5\text{–}25).$$

Substituting (6.5–23) for z^2 into (6.5–25) gives

$$\left(5y^2 + g_2 + 2y\sqrt{5y^2 + 3g_2}\right)\left(-4y^2 - 2g_2 + 2y\sqrt{5y^2 + 3g_2}\right)^2 = 108\, g_3^2$$
$$(6.5\text{–}26)$$

after multiplying both sides by 27. Now multiplying out the expressions on the left and dividing both sides by 4 gives

$$5y^6 - 2y^5\sqrt{5y^2 + 3g_2} + g_2^3 = 27g_3^2 \qquad (6.5\text{–}27).$$

Note that the powers y^n ($1 \leq n \leq 4$) outside the radical conveniently cancel during this process. Substituting the elliptic identity (6.4–30) into (6.5–27) and rearranging gives

$$5y^6 + \Delta = 2y^5\sqrt{5y^2 + 3g_2} \qquad (6.5\text{–}28)$$

which after squaring and combining terms gives

$$5y^{12} - 12g_2 y^{10} + 10\Delta y^6 + \Delta^2 = 0 \qquad (6.5\text{–}29).$$

The 12 roots of equation (6.5–29) are given by equation (6.5–14).

Perron[7] obtains the Jacobi sextic (6.4–35) from (6.5–29) by introducing the change of variable

$$y^2 = \frac{1}{s} \qquad (6.5\text{–}30)$$

and multiplying both sides by $\frac{s^6}{\Delta^2}$. Because of the nature of the transformation (6.5–30) and the roots (6.5–14) of equation 6.5–29, Perron[7] concludes that the roots of the Jacobi sextic (6.4–35) must be the Weierstrass elliptic functions

$$\sqrt{s_{mn}} = \frac{1}{\wp\left(\frac{4m\omega + 4n\omega'}{5}\right) - \wp\left(\frac{2m\omega + 2n\omega'}{5}\right)} \qquad (6.5\text{–}31).$$

6.6. Evaluation of the Weierstrass Elliptic Functions Using Genus 1 Theta Functions

The analysis up to this point provides a method for expressing the roots of a quintic equation by the expression (6.5–31) involving Weierstrass elliptic functions of the periods 2ω and $2\omega'$. The method for evaluating the Weierstrass elliptic functions (6.5–31) by means of theta functions of genus 1 as given by Kiepert[3] will now be discussed. Such theta functions (Section 4.3) use the argument q defined by

$$q = \exp\left(\frac{\pi i \omega'}{\omega}\right) \qquad (6.6\text{–}1).$$

Equation (6.6–1) indicates that q depends upon the period ratio ω'/ω of the elliptic functions. The rapidly converging series used to evaluate the relevant theta functions converge when $|q| < 1$.

In order to avoid a factor of the 24th root of unity, Kiepert[3] writes the six roots of the Jacobi sextic (6.4–1 or 6.4–35) as

$$\sqrt{s_\infty} = \frac{1}{\wp\left(\frac{2\omega}{5}\right) - \wp\left(\frac{4\omega}{5}\right)} \qquad (6.6\text{–}2)$$

$$\sqrt{s_k} = \frac{1}{\wp\left(\frac{2\omega' + 48k\omega}{5}\right) - \wp\left(\frac{4\omega' + 96k\omega}{5}\right)} \qquad (0 \le k \le 4) \qquad (6.6\text{–}3).$$

$$= \frac{1}{B} \sum_{i=-\infty}^{\infty} (-1)^i \varepsilon^{k(6i+1)^2} q^{(6i+1)^2/60} \qquad (6.6\text{--}26)$$

in which B is defined as in equation 6.6–20. The roots of the Jacobi sextic (6.4–1) can thus be expressed as quotients of two odd theta functions of genus 1 through (6.6–19), (6.6–20), and (6.6–26). There remains the problem of determining the periods 2ω, $2\omega'$, and therefore q through (6.6–1) corresponding to a given Jacobi sextic (6.4–1) defined by the parameters f and H corresponding to the vertex and face functions of the icosahedron used to generate the corresponding Brioschi quintic (6.3–7).

6.7. Evaluation of the Periods of the Elliptic Functions Corresponding to the Jacobi Sextic

Two methods are available for determining the periods corresponding to a Jacobi sextic (6.4–1):

(1) Use of hypergeometric series[9,10] based on the following differential equation:

$$J(1-J)\frac{d^2z}{dJ^2} + \left(\frac{2}{3} - \frac{7}{6}J\right) \cdot \frac{dz}{dJ} - \frac{z}{144} = 0 \qquad (6.7\text{--}1)$$

where

$$J = \frac{g_2^3}{\Delta} = \frac{V}{1728} \qquad (6.7\text{--}2)$$

The details of this method are beyond the scope of this book since they require consideration of the properties of hypergeometric series.

(2) Use of the Jacobi nome (Section 4.3). This method is the one that was used in the microcomputer program[4,5] and is discussed in greater detail in this section.

Determination of periods with the Jacobi nome uses the relationship between Jacobi and Weierstrass elliptic integrals (Section 4.1). Thus consider the Jacobi elliptic function[17]

---

[17] A. W. Erdelyi, W. Magnus, F. Oberhettinger, and F. G. Tricomi, *Higher Transcendental Functions*, McGraw-Hill, New York, 1953, Chapter 13.

$$w = \operatorname{sn} u \tag{6.7-3}$$

with modulus k defined by inversion of the integral

$$u = \int_0^w \frac{dx}{\sqrt{(1-x^2)(1-k^2x^2)}} \tag{6.7-4}$$

which has the square root of a special quartic polynomial in the denominator. Now write the differential equation 6.5–5 for the Weierstrass elliptic function in an analogous manner so that

$$w = \wp(u) \tag{6.7-5}$$

and

$$u = \int_0^w \frac{dx}{\sqrt{4x^3 - g_2 x - g_3}} = \int_0^w \frac{dx}{\sqrt{(x-e_1)(x-e_2)(x-e_3)}} \tag{6.7-6}.$$

The modulus k and complementary modulus k' can then be expressed[17] in terms of the roots e_1, e_2, and e_3 of the cubic polynomial of equation 6.7–6 by the relationships

$$k^2 = \frac{e_2 - e_3}{e_1 - e_3} \quad \text{and} \quad (k')^2 = 1 - k^2 = \frac{e_1 - e_2}{e_1 - e_3} \tag{6.7-7}.$$

Now calculate

$$L = \frac{1 - \sqrt{k'}}{1 + \sqrt{k'}} = \frac{\sqrt[4]{e_1 - e_3} - \sqrt[4]{e_1 - e_2}}{\sqrt[4]{e_1 - e_3} + \sqrt[4]{e_1 - e_2}} \tag{6.7-8}.$$

Then a value of q (equation 6.6–1) corresponding to a set of roots e_1, e_2, and e_3 of the cubic equation by using the Jacobi nome (Section 4.3)

$$q = \left(\frac{L}{2}\right) + 2\left(\frac{L}{2}\right)^5 + 12\left(\frac{L}{2}\right)^9 + 150\left(\frac{L}{2}\right)^{13} + \ldots = \sum_{j=1}^{\infty} q_j \left(\frac{L}{2}\right)^{4j+1} \quad (6.7\text{–}9)$$

in which the coefficients q_j form the series[8]

1; 2; 15; 150; 1707; 20,910; 268,616; 3,567,400; 48,555,069; 673,458,874; 9,481,557,398; 135,119,529,972; 1,944,997,539,623; 28,235,172,753,886

Thus the value of q corresponding to a particular Jacobi sextic can be obtained by the following sequence of steps:

(1) The cubic equation in (6.7–6) is solved after obtaining its coefficients g_2 and g_3 by means of equations 6.4–30, 6.4–40, and 6.4–41.

(2) The roots e_1, e_2, and e_3 of (6.7–6) are substituted into (6.7–8) to give L.

(3) The value obtained for L is substituted into the infinite series (6.7–9) to give q.

Equation 6.7–8 poses some difficulty since for a given ordering of the roots e_1, e_2, e_3 of the cubic equation in the denominator of equation 6.7–6, there are four fourth roots $\sqrt[4]{e_1 - e_3}$ and four fourth roots $\sqrt[4]{e_1 - e_2}$. The 16 possible combinations of these two fourth roots reduce to four possible values of L when the quotient (6.7–8) is taken. In addition, there are six permutations of the three roots e_1, e_2, and e_3 leading to $(4)(6) = 24$ possible q values for a given cubic equation. In general, half of these possible q values have $|q| > 1$ and therefore are unsatisfactory since the series for the corresponding theta functions do not converge. The computer can check the remaining 12 possible values of q as well as the apparent six roots of the Jacobi sextic by investigating whether the product

$$(s - s_\infty) \prod_{k=0}^{4} (s - s_k) \quad (6.7\text{–}10)$$

agrees with the original Jacobi sextic discarding any values of q with $|q| < 0$ which give incorrect roots. The following ambiguities also arise after substituting the value for q obtained from (6.7–9) into the theta series formulas 6.6–20, 6.6–21, and 6.6–26:

(1) The three values of $\left(\sqrt[6]{\Delta}\right)^2$ (see equations 6.6–20 and 6.6–21) differing by factors of $\exp(2\pi i/3)$ need to be checked.

(2) The q factor in equation (6.6–26) is obtained by using the fifth root of q in an algorithm (i.e., the 60 in the denominator of the exponent in equation 6.6–26) to determine the infinite sum in the equation (6.6–20) for B. Thus all five of the fifth roots of q must be checked.

(3) These 15 possible choices of $\sqrt[5]{q}$ and $\left(\sqrt[6]{\Delta}\right)^2$ are all used in the computer algorithm[4,5] and the apparent six roots of the Jacobi sextic have been checked using (6.7–10). Choices giving incorrect roots were rejected.

6.8. Undoing the Tschirnhausen Transformations

Equations 6.6–20, 6.6–21, and 6.6–26 express the six roots of the Jacobi sextic (6.4–35) in terms of genus 1 theta functions expressed as rapidly converging infinite sums. In order to convert these roots to the roots of the original general quintic (6.2–1), the various Tschirnhausen and other transformations need to be undone in the following sequence:

s_∞, s_k	$\longrightarrow$	y_k	$\longrightarrow$	z_k	$\longrightarrow$	x_k
Jacobi sextic		Brioschi quintic		Principal quintic		General quintic
(6.4–35)		(6.3–1)		(6.2–2)		(6.2–1)

In order to obtain the roots y_k of the Brioschi quintic (6.3–1) from the roots of the Jacobi sextic (6.4–36), the following two equations are used in order to assure the correct selection of the sign for the outer square root in (6.4–5):

$$y_k^2 = \frac{1}{\sqrt{5}}(s_\infty - s_k)(s_{k+2} - s_{k+3})(s_{k+4} - s_{k+1}) \qquad (6.8\text{–}1)$$

The Kiepert Algorithm for the Quintic Equation

$$y_k = \frac{\frac{216 g_3}{\Delta^3}}{(y_k^2)^2 + \frac{10}{\Delta} y_k^2 + \frac{45}{\Delta^2}} \tag{6.8-2}$$

The roots z_k of the principal quintic are then obtained from the roots y_k of the Brioschi quintic by using (6.3–2). Finally the roots x_k of the general quintic (6.2–1) are obtained from the roots z_k of the principal quintic (6.2–2) by the equation

$$x_k = -\frac{E + (z_k - v)(u^3 + Au^2 + Bu + C) + (z_k - v)^2(2u + A)}{u^4 + Au^3 + Bu^2 + Cu + D + (z_k - v)(3u^2 + 2Au + B) + (z_k - v)^2} \tag{6.8-3}$$

In order to derive equation (6.8–3) first write equation (6.2–4) as

$$(x - u)^2 = (z - v) - u(x - u) \tag{6.8-4}.$$

Iteratively, find equations of the form

$$(x - u)^m = P_m(u, z - v) + Q_m(u, z - v)(x - u) \tag{6.8-5}$$

for $3 \leq m \leq 5$ where P_m and Q_m are polynomials. Substituting these equations (6.8–5) into (6.2–11) leads ultimately to a linear equation in $(x - u)$, which when solved and simplified yields the desired equation 6.8–3.

Chapter 7

The Methods of Hermite and Gordan for Solving the General Quintic Equation

7.1 Hermite's Early Work on the Quintic Equation

The previous chapter discusses in detail the Kiepert algorithm for solving the general quintic equation since this is the method that we have tested on the computer. This chapter summarizes the earlier methods of Hermite and Gordan for solving the general quintic equation starting with the original 1858 work of Hermite[1] in this area.

Hermite first looked at the cubic equation

$$x^3 - 3x + 2a = 0 \qquad (7.1-1)$$

in which the constant term a is represented by the sine of an angle α so that the roots of the equation separate into the three functions

$$2\sin\frac{\alpha}{3}, \qquad 2\sin\frac{\alpha+2\pi}{3}, \qquad 2\sin\frac{\alpha+4\pi}{3} \qquad (7.1-2)$$

He then considered an extension of this approach to the Bring-Jerrard form of the quintic equation

$$x^5 - x - a = 0 \qquad (7.1-3)$$

realizing that the general quintic can be transformed to the Bring-Jerrard quintic by solving equations of degrees no higher than three. In this connection Hermite[1] recognized the role that elliptic transcendents can play in the solution of the Bring-Jerrard quintic (7.1-3) analogous to the role of the trigonometric functions in the solution of the cubic equation (7.1-1) (see Section 5.1).

Let K and K' be the periods of the elliptic integral

$$\int \frac{d\varphi}{\sqrt{1 - k^2\sin^2\varphi}} \qquad (7.1-4)$$

i.e.,

---

[1] C. Hermite, Sur la Résolution de l'Équation du Cinquième Degré, *Compt. Rend.*, **46**, 508–515 (1858).

Other Methods for Solving the General Quintic Equation

$$K = \int_0^{\pi/2} \frac{d\varphi}{\sqrt{1 - k^2\sin^2\varphi}} \quad \text{and} \quad K' = \int_0^{\pi/2} \frac{d\varphi}{\sqrt{1 - (k')^2\sin^2\varphi}} \quad (7.1\text{-}5)$$

In addition

$$q = \exp\left(-\pi\frac{K'}{K}\right) \quad \text{(compare equation 6.6–1)} \quad (7.1\text{-}6a)$$

and $\quad k^2 + (k')^2 = 1 \quad (7.1\text{-}6b)$.

The fourth root of the modulus k is related to q by the infinite series

$$\sqrt[4]{k} = \sqrt{2}\sqrt[8]{q}\frac{\sum q^{2m^2+m}}{\sum q^{m^2}} \quad (7.1\text{-}7)$$

Define the following:

$$q = e^{i\pi\tau} \quad (7.1\text{-}8a)$$

$$\sqrt[4]{k} = \varphi(\tau) = \sqrt{\frac{\theta_2(0)}{\theta_3(0)}} \quad \text{(compare equation 4.3–14a)} \quad (7.1\text{-}8b)$$

$$\sqrt[4]{k'} = \psi(\tau) = \sqrt{\frac{\theta_4(0)}{\theta_3(0)}} \quad \text{(compare equation 4.3–14b)} \quad (7.1\text{-}8c)$$

The variable $\tau = \omega'/\omega$ is the same as in Section 4.3 but is designated as ω by Hermite[1] and 19th century authors discussing Hermite's work.

The functions φ and ψ have the following properties:

$$\varphi^8(\tau) + \psi^8(\tau) = 1 \text{ (because } k^2 + (k')^2 = 1) \quad (7.1\text{-}9a)$$

$$\varphi\left(\frac{1}{\tau}\right) = \psi(\tau) \quad (7.1\text{-}9b)$$

$$\varphi(\tau+1) = \exp\left(\frac{i\pi}{8}\right)\frac{\varphi(\tau)}{\psi(\tau)} \quad (7.1\text{-}9c)$$

$$\psi(\tau+1) = \frac{1}{\psi(\tau)} \quad (7.1\text{-}9d)$$

Now let n be a prime number and define u and v as follows:

$$v = \varphi(n\tau) \text{ and } u = \varphi(\tau) \qquad (7.1\text{--}10)$$

The parameters u and v are linked by an equation of degree $n + 1$ known as the *modular equation*, the roots of which have some special properties. Now let ε be 1 or -1 depending on whether 2 is a quadratic residue or not a quadratic residue with respect to n, respectively. The $n + 1$ roots u of the modular equation have the form

$$\varepsilon\varphi(n\tau) \text{ and } \varphi\left(\frac{\tau+16m}{n}\right) \qquad (7.1\text{--}11)$$

where m is an integer taken modulo n. Thus the modular equation of the sixth degree (i.e., $n = 5$) has the following form:

$$u^6 - v^6 + 5u^2v^2(u^2 - v^2) + 4uv(1 - u^4v^4) = 0 \qquad (7.1\text{--}12)$$

whose six roots have the form 7.1–11 where $n = 5$ and $m = 0, 1, 2, 3,$ and 4.

The modular equation of the sixth degree (7.1–12) can be related to the Bring-Jerrard quintic (7.1–3). Thus consider the following function of the six roots of equation 7.1–12:

$$\Phi(\tau) = \left[\varphi(5\tau) + \varphi\left(\frac{\tau}{5}\right)\right]\left[\varphi\left(\frac{\tau+16}{5}\right) - \varphi\left(\frac{\tau+64}{5}\right)\right]\left[\varphi\left(\frac{\tau+32}{5}\right) - \varphi\left(\frac{\tau+48}{5}\right)\right]$$
$$(7.1\text{--}13)$$

The five quantities $\Phi(\tau)$, $\Phi(\tau + 16)$, $\Phi(\tau + 32)$, $\Phi(\tau + 48)$, and $\Phi(\tau + 64)$ are the roots of a quintic equation of which the coefficients are rational in $\varphi(\tau)$, namely

$$\Phi^5 - 2000\, \varphi^4(\tau)\psi^{16}(\tau)\Phi - 1600\sqrt{5}\, \varphi^3(\tau)\psi^{16}(\tau)[1 + \varphi^8(\tau)] = 0 \qquad (7.1\text{--}14).$$

This can be converted to the Bring-Jerrard quintic (7.1–3) by the substitution

$$\Phi = \sqrt[4]{2^4 5^3}\, \varphi(\tau)\psi^4(\tau)x \qquad (7.1\text{--}15)$$

leading to

$$x^5 - x - \left(\frac{2(1+\varphi^8(\tau))}{\sqrt[4]{5^5\varphi^2(\tau)\psi^4(\tau)}}\right) = 0 \qquad (7.1\text{--}16).$$

Thus in order to determine the roots of the Bring-Jerrard quintic (7.1–3) by the function $\Phi(\tau)$, it is necessary to determine τ or rather $\varphi(\tau)$ corresponding to the following:

$$a = \frac{2(1+\varphi^8(\tau))}{\sqrt[4]{5^5 \varphi^2(\tau) \psi^4(\tau)}} \qquad (7.1\text{–}17)$$

This relationship is the basis of Hermite's method for solving the general quintic equation.

The initial step of Hermite's method is the Tschirnhausen transformation of the general quintic equation to the Bring-Jerrard quintic (7.1–3), a process which is difficult (Section 3.3) although it requires the solution of equations of degrees no higher than three.[2] A parameter A is then defined by the following equation:

$$A = \frac{a\sqrt[4]{5^5}}{2} \qquad (7.1\text{–}18)$$

The corresponding elliptic modulus k (equations 7.1–4 and 7.1–5) can be obtained by solving the quartic equation

$$k^4 + A^2 k^3 + 2k^2 - A^2 k + 1 = 0 \qquad (7.1\text{–}19).$$

Let $\sin \alpha = \dfrac{1}{4A^2}$ \qquad (7.1–20).

Then the roots of equation 7.1–19 are the following:

$$k = \tan\frac{\alpha}{4},\ \tan\frac{\alpha+2\pi}{4},\ \tan\frac{\pi-\alpha}{4},\ \tan\frac{3\pi-\alpha}{4} \qquad (7.1\text{–}21)$$

The roots of the Bring-Jerrard quintic are then determined by the following equation ($i = 0, 1, 2, 3, 4$):

---

[2] A. Cayley, On Tschirnhausen's Transformation, *Phil Trans. Roy. Soc. London*, **152**, 561–578 (1862); *Collected Mathematical Papers*, Cambridge Univ. Press, 1891, Vol IV, pp. 375–394.

$$x_i = \frac{\Phi(\tau + 16i)}{\sqrt[4]{2^4 5^3 \varphi(\tau) \psi^4(\tau)}} \qquad (7.1\text{–}22)$$

This formula is a more complicated analogue of the formula 5.1–10b in Section 5.1 for solution of the general cubic equation using trigonometric functions.

Hermite also considered the problem of the numerical calculation of these roots noting that the infinite series defining $\varphi(\tau)$ and $\psi(\tau)$ (equations 7.1–7) converge extremely rapidly even if q has an imaginary component (compare Section 4.3). The quantities $\Phi(\tau)$, $\Phi^2(\tau)$, and $\Phi^3(\tau)$ can be estimated by the following infinite series in which $Q = q^{1/5}$:

$$\Phi(\tau) = \sqrt{2^3 5}\ \sqrt[8]{Q^3}\ (1 + Q - Q^2 + Q^3 - 8Q^5 - 9Q^6 + \ldots) \qquad (7.1\text{–}23a)$$

$$\Phi^2(\tau) = 2^3 5 \sqrt[4]{Q^3}\ (1 + 2Q - Q^2 + 3Q^4 - 18Q^5 - 33Q^6 + 14Q^7 + \ldots) \qquad (7.1\text{–}23b)$$

$$\Phi^3(\tau) = \sqrt{2^9 5^3}\ \sqrt[8]{Q^9}\ (1 + 3Q - 2Q^3 + 6Q^4 - 24Q^5 - 79Q^6 + \ldots) \qquad (7.1\text{–}23c)$$

Note that the series for $\Phi(\tau)$, $\Phi^2(\tau)$, and $\Phi^3(\tau)$ are missing the powers of Q of which the exponents mod 5 are 4, 3, and 2, respectively. The effect of changing τ to $\tau + 16m$ in the arguments of Φ has the effect of multiplying Q by the diverse fifth roots of unity.

7.2 Gordan's Work on the Quintic Equation

The algorithm by Hermite[1] for solution of the general quintic equation requires the difficult Tschirnhausen transformation of the general quintic equation to the Bring-Jerrard quintic equation (7.1–3). The Kiepert algorithm avoids this difficulty by using the properties of polyhedral functions to transform the principal quintic equation

$$z^5 + 5az^2 + 5bz + c = 0 \qquad (7.2\text{–}1)$$

to the one-parameter Brioschi quintic equation

$$y^5 - 10Zy^3 + 45Z^2y - Z^2 = 0 \qquad (7.2\text{–}2)$$

Other Methods for Solving the General Quintic Equation

which can then be solved using elliptic functions (Chapter 6). In addition Gordan[3] described in 1878 an alternative method based on invariant theory (Section 2.6)[4,5] for solving the principal quintic equation which also avoids the difficult Tschirnhausen transformation to the Bring-Jerrard quintic equation.

In order to set up his method Gordan[3] represents the 120 symmetry operations of the icosahedron by the transformations in Table 7–1 of the homogeneous variables y_1 and y_2 (corresponding to u and v in Section 2.5 so that z of the Riemann sphere is y_1/y_2). The icosahedral functions (compare Section 2.5)

$$f = y_1 y_2 (y_1^{10} + 11 y_1^5 y_2^5 - y_2^{10}) \tag{7.2–3a}$$

$$T = y_1^{30} + 522 y_1^{25} y_2^5 - 10005 y_1^{20} y_2^{10} - 10005 y_1^{10} y_2^{20} - 522 y_1^5 y_2^{25} + y_2^{30} \tag{7.2–3b}$$

$$H = -y_1^{20} + 228 y_1^{15} y_2^5 - 494 y_1^{10} y_2^{10} - 228 y_1^5 y_2^{15} - y_2^{20} \tag{7.2–3c}$$

are not altered by the 120 permutations in Table 7–1.

Table 7–1: The 120 Icosahedral Permutations of the Homogeneous Variables y_1 and y_2

$[\mu, \nu = 0, 1, 2, 3, \text{ and } 4; \; \varepsilon = \cos \frac{2\pi}{5} + i \sin \frac{2\pi}{5}]$

y_1	y_2
$\pm \varepsilon^{\nu/2} y_1$	$\pm \varepsilon^{-\nu/2} y_2$
$\mp \varepsilon^{\nu/2} y_2$	$\pm \varepsilon^{-\nu/2} y_1$
$\pm \varepsilon^{\mu/2} \dfrac{[(\varepsilon+\varepsilon^4)\varepsilon^{-\nu/2} y_1 + \varepsilon^{\nu/2} y_2]}{\varepsilon^2 - \varepsilon^3}$	$\pm \dfrac{\varepsilon^{-\mu/2}[\varepsilon^{-\nu/2} y_1 - (\varepsilon+\varepsilon^4)\varepsilon^{\nu/2} y_2]}{\varepsilon^2 - \varepsilon^3}$
$\mp \varepsilon^{\mu/2} \dfrac{[\varepsilon^{-\nu/2} y_1 - (\varepsilon+\varepsilon^4)\varepsilon^{\nu/2} y_2]}{\varepsilon^2 - \varepsilon^3}$	$\pm \dfrac{\varepsilon^{-\mu/2}[(\varepsilon+\varepsilon^4)\varepsilon^{-\nu/2} y_1 + \varepsilon^{\nu/2} y_2]}{\varepsilon^2 - \varepsilon^3}$

---

[3] P. Gordan, Über die Auflösung der Gleichungen vom fünften Grade, *Math. Ann.*, **13**, 375–404 (1878).
[4] J. H. Grace and A. Young, *The Algebra of Invariants*, Cambridge, 1903.
[5] O. E. Glenn, *A Treatise on the Theory of Invariants*, Ginn and Co., Boston, 1915.

The substitutions in Table 7–1 also form an icosahedral group if a complex root of unity other than ε is used. If ε^4 is used, then the new substitution system goes back to the original one if $-y_2$ and y_1 are substituted for y_1 and y_2, respectively. However, use of ε^2 generates a new set of substitutions represented in Table 7–2 by the variables x_1 and x_2.

Table 7–2: The 120 Icosahedral Permutations of the Homogeneous Variables x_1 and x_2

$[\mu,\nu = 0, 1, 2, 3, \text{ and } 4; \varepsilon = \cos\frac{2\pi}{5} + i\sin\frac{2\pi}{5}]$

x_1	x_2
$\pm \varepsilon^\nu x_1$	$\pm \varepsilon^{-\nu} x_2$
$\mp \varepsilon^\nu x_2$	$\pm \varepsilon^{-\nu} x_1$
$\pm\dfrac{\varepsilon^\mu[(\varepsilon^2+\varepsilon^3)\varepsilon^{-\nu}x_1+\varepsilon^\nu x_2]}{\varepsilon^4-\varepsilon}$	$\pm\dfrac{\varepsilon^{-\mu}[\varepsilon^{-\nu}x_1-(\varepsilon^2+\varepsilon^3)\varepsilon^\nu x_2]}{\varepsilon^4-\varepsilon}$
$\pm\dfrac{\varepsilon^\mu[\varepsilon^{-\nu}x_1-(\varepsilon^2+\varepsilon^3)\varepsilon^\nu x_2]}{\varepsilon^4-\varepsilon}$	$\pm\dfrac{\varepsilon^{-\mu}[(\varepsilon^2+\varepsilon^3)\varepsilon^{-\nu}x_1+\varepsilon^\nu x_2]}{\varepsilon^4-\varepsilon}$

Gordan[3] now seeks the lowest degree homogeneous entire function of y_1, y_2, x_1, x_2 which remains unchanged when y_1 and y_2 are subject to the permutations in Table 7–1 and x_1 and x_2 are subject to the permutations in Table 7–2. This function turns out to be

$$\xi = y_1^3 x_1^2 x_2 + y_1^2 y_2 x_2^3 + y_1 y_2^2 x_1^3 - y_2^3 x_1 x_2^2 \qquad (7.2\text{–}4)$$

This function is designated in this book by ξ rather than the f used by Gordan to avoid confusion with the f used for the icosahedral vertex function (e.g., equation 7.2–3a).

Gordan next derives a system of covariants of ξ by using a generalized version of the transvection process (Section 2.6) leading to the following functions in x_1, x_2, y_1, y_2:

Degree 8:
$$\varphi = {}^9\!/_4(\xi,\xi)_{1,1} = y_1^4 x_1 x_2^3 - y_1^3 y_2 x_1^4 - 3y_1^2 y_2^2 x_1^2 x_2^2 + y_1 y_2^3 x_2^4 + y_2^4 x_1^3 x_2 \tag{7.2-5a}$$

Degree 10
$$\psi = 12(\xi,\varphi)_{1,1} = y_1^5(x_1^5 + x_2^5) - 10 y_1^4 y_2 x_1^3 x_2^2 + 10 y_1^3 y_2^2 x_1 x_2^4 +$$
$$10 y_1^2 y_2^3 x_1^4 x_2 + 10 y_1 y_2^4 x_1^2 x_2^3 + y_2^5(-x_1^5 + x_2^5) \tag{7.2-5b}$$

The functions φ and ψ, like the function ξ from which they are obtained by the indicated transvections, are unaltered by the 120 icosahedral substitutions (Tables 7–1 and 7–2).

Gordan[3] now uses these functions of icosahedral symmetry as the coefficients of a principal quintic equation as follows:
$$\chi^5 + 5\xi\chi^2 - 5\varphi\chi - \psi = 0 \tag{7.2-6}$$
A root of this equation can be expressed in the following form:
$$\chi = -y_1(x_1 + x_2) + y_2(x_1 - x_2) \tag{7.2-7}$$
The other roots then have the form
$$\chi_\nu = -\varepsilon^\nu x_1 y_1 + \varepsilon^{2\nu} x_1 y_2 - \varepsilon^{3\nu} x_2 y_1 - \varepsilon^{4\nu} x_2 y_2 \tag{7.2-8}$$
where ε is a complex fifth root of unity.

The roots of equation (7.2–6) can also be related to the icosahedral polynomials f, H, and T (equations 7.2–3) as well as the polynomials t_ν and W_ν of the five suboctahedra of the icosahedron defined as the midpoints of its edges (compare Section 6.3) by the following polynomials in x_1 and x_2 (or analogously y_1 and y_2):

$$t_\nu = \varepsilon^{3\nu} x_1^6 + 2\varepsilon^{2\nu} x_1^5 x_2 - 5\varepsilon^\nu x_1^4 x_2^2 - 5\varepsilon^{4\nu} x_1^2 x_2^4 - 2\varepsilon^{3\nu} x_1 x_2^5 + \varepsilon^{2\nu} x_2^6 \tag{7.2-9a}$$

$$W_\nu = -\varepsilon^{4\nu} x_1^8 + \varepsilon^{3\nu} x_1^7 x_2 - 7\varepsilon^{2\nu} x_1^6 x_2^2 - 7\varepsilon^\nu x_1^5 x_2^3 + 7\varepsilon^{4\nu} x_1^3 x_2^5 -$$
$$7\varepsilon^{3\nu} x_1^2 x_2^6 - \varepsilon^{2\nu} x_1 x_2^7 - \varepsilon^\nu x_2^8 \tag{7.2-9b}$$

The parameter c is obtained by solving the following quadratic equation:

$$(\xi\psi - 8\varphi^2)c^2 - (9\xi^2\varphi - {}^1\!/_3\psi^2)c + \frac{(-27\xi^4 - 8\varphi^3 + 9\xi\varphi\psi)}{9} = 0 \tag{7.2-10}$$

Equation 7.2–10 plays an analogous role in the Gordan algorithm to the quadratic equation in λ (equation 6.3–28) used in the Kiepert algorithm. The parameter c is the value of a function that can be derived from ξ and φ by further transvection processes and thus is invariant under icosahedral substitutions (Tables 7–1 and 7–2). The quantities ξ, φ, ψ, and c are then used to calculate the parameters A, B, C, and D by the following equations:

$$A = -\frac{8\xi\varphi + 3c\psi}{\xi^2 + 3c\varphi} \qquad (7.2\text{–}11\text{a})$$

$$B = \frac{-8\varphi^2 + \xi\psi}{\xi^2 + 3c\varphi} \qquad (7.2\text{–}11\text{b})$$

$$C = -\frac{-8\varphi^3 + 9\xi\varphi\psi + 3c\psi^2}{\xi^2 + 3c\varphi} \qquad (7.2\text{–}11\text{c})$$

$$D = \frac{AB^2 + 3cAB + 30\xi B}{\varphi} \qquad (7.2\text{–}11\text{d})$$

The roots of the principal quintic (7.2–8) have the following form in terms of the polyhedral functions of the icosahedron and its five suboctahedra:

$$\chi_v = -\left(\frac{Af}{H}\right)W_v - \left(\frac{Df^3}{HT}\right)t_v W_v \qquad (7.2\text{–}12)$$

Note the similarity of equation 7.2–12 to the analogous equation (6.3–8) in the Kiepert algorithm.

Gordan[3] obtains an equation analogous to the Brioschi quintic (6.3–7) by calculating its variable u from the equation

$$u = -\sqrt{B}\,\frac{3\chi + AZ}{A\chi + BZ} \qquad (7.2\text{–}13)$$

Other Methods for Solving the General Quintic Equation   137

An analogue in Gordan's method to the Brioschi quintic of Kiepert's algorithm then has the form

$$u^5 - 10u^3 + 45u - \frac{H}{\sqrt{f^5}} = 0 \qquad (7.2\text{--}14)$$

Equation (7.2–14) can then be related to the Brioschi quintic (6.3–7) by the change of variables $t = u\sqrt{f}$.

Gordan's method for solution of the general quintic equation thus first uses the standard Tschirnhausen transformation for conversion of the general quintic equation to the principal quintic (Section 6.2), which is now expressed as equation (7.2–6). His method can then deliver a Brioschi-like quintic (7.2–14) similar to the portion of the Kiepert algorithm discussed in Section 6.3. In order to obtain the actual roots of the principal quintic (7.2–6) the elliptic modulus k^2 is calculated from the following algebraically soluble sextic equation in k^2:

$$-\frac{16(1 + 14k^2 + k^4)^3}{k^2(1 - k^2)^4} = \frac{(A^2 - 3B)^3}{C} \qquad (7.2\text{--}15)$$

The solution of equation 7.2–15 to determine the elliptic modulus k^2 requires solution of a cubic equation just like the Kiepert algorithm requires solution of the cubic equation in equation (6.7–6), namely

$$4x^3 - g_2 x - g_3 = 0 \qquad (7.2\text{--}16)$$

from which the modulus k^2 is obtained by the quotient

$$k^2 = \frac{e_2 - e_3}{e_1 - e_3} \qquad (7.2\text{--}17)$$

where e_1, e_2, and e_3 are the three roots of the cubic equation 7.2–16.

The remainder of Gordan's solution of the quintic equation uses many of the methods of Hermite (Section 7.1). Thus the following relationship is used to obtain q and more significantly ω:

$$\sqrt[4]{k} = \varphi(\omega) = \sqrt{2} \sqrt[8]{q} \frac{\sum q^{2m^2+m}}{\sum q^{m^2}} \tag{7.2-18}$$

Then the roots of a Bring-Jerrard like quintic are obtained from the equation derived by Hermite, namely

$$h_v = 2\sqrt{5} \sqrt[4]{k} \left[\varphi(5\omega) + \varphi\left(\frac{\omega+16v}{5}\right) \right]$$
$$\times \left[\varphi\left(\frac{\omega+16(v+1)}{5}\right) - \varphi\left(\frac{\omega+16(v+4)}{5}\right) \right]$$
$$\times \left[\varphi\left(\frac{\omega+16(v+2)}{5}\right) - \varphi\left(\frac{\omega+16(v+3)}{5}\right) \right] \tag{7.2-19}$$

The roots of the original principal quintic (7.2–6) can be obtained by substituting the roots h_v and the elliptic moduli k and k' into equation 7.2–12 leading to

$$\chi_v = -\frac{Ak}{4(1+14k^2+k^4)} \left[\frac{2(1+k^2)h_v}{\sqrt{k}} + \frac{4\sqrt{k}\,(h_v^3+2kh_v)}{2h_v\sqrt{k}+k^2+1} \right]$$
$$- \frac{Dk^2(1-k^2)^2}{4(1+14k^2+k^4)(1-34k^2+k^4)} \left[\frac{3\sqrt{k}}{1+k^2} - \frac{h_v^3+2kh_v}{2\sqrt{k}\,(2h_v\sqrt{k}+k^2+1)} \right] \tag{7.2-20}$$

The roots of the general quintic equation can then be determined from the roots of the principal quintic (7.2–20) by reversing the original Tschirnhausen transformation from the general quintic to the principal quintic by equation (6.8–3).

Chapter 8

Beyond the Quintic Equation

8.1 The Sextic Equation

The theory of the general sextic (sixth degree) equation was discussed by Cole[1] in 1886 within a decade of the publication of the Kiepert[2] and Gordan[3] algorithms for solution of the quintic equation. This theory extends ideas found to be useful for equations of degrees 5 and lower to the sextic equation.

A central feature of the theories of equations one to five is the role played played by certain groups of linear transformations of a single variable corresponding to the Galois groups of the equations. The quartic equation is the lowest degree equation in which the group of linear transformations plays a role, which, however, is not apparent in the method for solution of the quartic equation discussed in Section 5.2. The four roots of the quartic equation have six anharmonic ratios, namely λ, $\frac{1}{\lambda}$, $1 - \lambda$, $\frac{1}{1 - \lambda}$, $\frac{\lambda - 1}{\lambda}$, and $\frac{\lambda}{\lambda - 1}$ as discussed in Section 4.1 (see particularly equations 4.1–13 to 4.1–19). These six anharmonic ratios are roots of a sextic equation, which is not the general sextic equation but is characterized by the fact that every root is a rational linear fraction of every other root. This sextic equation reduces to the resolvent cubic equation (5.2–6) found in the solution of the quartic equation discussed in Section 5.2.

The corresponding group of linear transformations $z \rightarrow z'$ for the general quintic equation corresponds to the 60 operations of the alternating

---

[1] Cole, F. N., A Contribution to the Theory of the General Equation of the Sixth Degree, *Amer. J. Math.*, **8**, 265–286 (1886).
[2] L. Kiepert, Auflösung der Gleichungen fünften Grades, *J. für Math.*, **87**, 114–133 (1878).
[3] P. Gordan, Über die Auflösung der Gleichungen vom fünften Grade, *Math. Ann.*, **13**, 375–404 (1878).

group A_5, which can be represented by the following equation where z is an integer congruent to 0, 1, 2, 3, 4, or ∞ mod 5:

$$z' \equiv \frac{\alpha z + \beta}{\gamma z + d} \qquad (8.1\text{–}1)$$

By proper choice of α, β, γ, and δ, z' is an integer. It turns out that there are 60 sets of values for α, β, γ, and δ that satisfy these conditions. Selection of any of these sets and putting z successively congruent to 0, 1,..., 4, and ∞ mod 5 leads to z' values congruent to 0,...,∞ but in a different order from the original. Thus a rational function of the roots of the quintic equation can be found which, when these roots undergo any *even* permutation, is linearly transformed by the formula

$$\phi' = \frac{\alpha \phi + \beta}{\gamma \phi + d} \qquad (8.1\text{–}2)$$

The function ϕ plays an analogous role in the theory of the quintic equation to the anharmonic ratio λ in the theory of the quartic equation. The function ϕ satisfies an equation of degree 60, which is related to the icosahedral polynomials (Section 2.5), which are involved in the solution of the general quintic equation (Sections 6.3 and 7.2). There is the important difference that the equation in λ is simpler than the general quartic equation, i.e., is reducible to a cubic and quadratic equation, whereas the equation in ϕ still has the same A_5 Galois group of the original quintic equation.

Extension of such ideas to the general sextic equation requires finding a group of linear transformations isomorphic with the full symmetric group S_6 with 6! = 720 operations or the corresponding alternating group A_6 with 6!/2 = 360 operations. Klein[4] has shown that there is no such finite group of linear transformations of a single variable. The smallest number of variables to represent the symmetric S_6 group through linear substitutions is three, which becomes four if the linear transformation is written in homogeneous form. Of

---

[4]F. Klein, *Vorlesungen über das Ikosaeder*, Teubner, Leipzig, 1884.

these four homogeneous variables, the ratio of three to the fourth will then be transformed by a nonhomogeneous linear transformation. This group of transformations is known under the geometrical form in which this theory has been treated in connection with a surface of fourth order and class known as Kummer's surface.[5]

The inability to represent the S_6 Galois group of the general sextic equation as a linear transformation of a single variable is a new complication that arises when the degree of an algebraic equation increases from 5 to 6. A consequence of this is that the elliptic functions derived from elliptic integrals of the type

$$f(x) = \int \frac{dx}{\sqrt{\mathcal{P}(x)}} \qquad (8.1\text{--}3)$$

in which $\mathcal{P}(x)$ is a polynomial of degree 3 or 4 (Chapter 4) are not sufficient for the solution of general equations of degrees 6 or higher. Hyperelliptic integrals of the type (8.1–3) in which $\mathcal{P}(x)$ is a polynomial of degree 5 or 6 are necessary for solution of the general sextic equation. These integrals are related to double theta functions (Section 4.4).

Bolza[6] and Maschke[7] have published back-to-back papers in *Mathematische Annalen* that provide a basis for solving the general sextic equation using double theta functions. Solution of the general sextic equation by this method is also discussed by Brioschi[8] along with the covariants and invariants of the general sextic equation.

The general sextic equation

$$x^6 + a_1 x^5 + a_2 x^4 + a_3 x^3 + a_4 x^2 + a_5 x + a_6 = 0 \qquad (8.1\text{--}4)$$

---

[5] R. W. H. Hudson, *Kummer's Quartic Surface*, Cambridge, 1905.
[6] O. Bolza, Darstellung der Rationalen ganzen Invarianten der Binärform sechsten Grades durch die Nullwerthe der zugehörigen θ-Functionen, *Math. Ann.*, **30**, 478–495 (1887).
[7] H. Maschke, Über die quaternäre, endliche, lineare Substitutionsgruppe der Borschardt'schen Moduln, *Math. Ann.*, **30**, 496–515 (1887).
[8] F. Brioschi, Sur l'Équation du Sixième Degré, *Acta Math.*, **12**, 83–101 (1888).

can be converted to the special Maschke sextic equation

$$\begin{aligned}\Phi(y) &= y^6 - 6F_8 y^4 + 4F_{12} y^3 + 9F_8^2 y^2 - 12F_{20} y + 4F_{24} \\ &= (9y^3 - 3F_8 y + 2F_{12})^2 + 12(F_8 F_{12} - F_{20})y - 4(F_{12}^2 - F_{24}) \\ &= 0 \end{aligned} \quad (8.1\text{–}5)$$

using a Tschirnhausen transformation requiring solution of a quartic equation so that the necessary relationship can be found between the coefficients of y^4 and y^2. The coefficients F_8, F_{12}, F_{20}, and F_{24} determine the invariants of a binary sextic $S(x_1, x_2)$. The periods τ_{11}, τ_{12}, and τ_{22} from the corresponding hyperelliptic integral

$$u(a) = \int_1^a \frac{d(x_1/x_2)}{\sqrt{S(x_1/x_2, 1)}} \quad (8.1\text{–}6)$$

are substituted into the following four even double theta functions of zero argument (Section 4.4) to give the Borchardt moduli

$$z_1 = \theta \binom{00}{11}(0|2\tau_{11}, 2\tau_{12}, 2\tau_{22}) \quad (8.1\text{–}7a)$$

$$z_2 = \theta \binom{11}{00}(0|2\tau_{11}, 2\tau_{12}, 2\tau_{22}) \quad (8.1\text{–}7b)$$

$$z_3 = \theta \binom{10}{00}(0|2\tau_{11}, 2\tau_{12}, 2\tau_{22}) \quad (8.1\text{–}7c)$$

$$z_4 = \theta \binom{01}{11}(0|2\tau_{11}, 2\tau_{12}, 2\tau_{22}) \quad (8.1\text{–}7d)$$

These Borchardt moduli provide the equation of a Kummer surface[5]

$$\phi + A\psi_2 + B\psi_3 + C\psi_4 + 2D\chi = 0 \quad (8.1\text{–}8)$$

where

$$\phi = z_1^4 + z_2^4 + z_3^4 + z_4^4 \quad (8.1\text{–}9a)$$

$$\psi_2 = z_1^2 z_2^2 + z_3^2 z_4^2 \quad (8.1\text{–}9b)$$

$$\psi_3 + z_1^2 z_3^2 + z_2^2 z_4^2 \quad (8.1\text{–}9c)$$

$$\psi_4 = z_1^2 z_4^2 + z_2^2 z_3^2 \quad (8.1\text{–}9d)$$

$$\chi = z_1 z_2 z_3 z_4 \quad (8.1\text{–}9e)$$

The roots of the Maschke sextic (8.1–5) are then

$$y_1 = \phi + 6(-\psi_2 - \psi_3 - \psi_4) \qquad (8.1\text{--}10a)$$
$$y_2 = \phi + 6(-\psi_2 + \psi_3 + \psi_4) \qquad (8.1\text{--}10b)$$
$$y_3 = \phi + 6(\psi_2 - \psi_3 + \psi_4) \qquad (8.1\text{--}10c)$$
$$y_4 = \phi + 6(\psi_2 + \psi_3 - \psi_4) \qquad (8.1\text{--}10d)$$
$$y_5 = -2\phi - 24\chi \qquad (8.1\text{--}10e)$$
$$y_6 = -2\phi + 24\chi \qquad (8.1\text{--}10f)$$

Reversing the Tschirnhausen transformation converting the general sextic equation (8.1–4) to the Maschke sextic equation (8.1–5) can then give the roots of the original general sextic equation (8.1–4).

8.2 The Septic Equation

Table 2–3 lists seven possible transitive permutation groups of degree 7. All of these groups are possible Galois groups for irreducible septic equations (equations of degree 7). The most interesting group in this list is the group designated as $L(3,2)$ of order 168. This is the smallest simple group which is not an alternating group, A_n ($n \geq 5$). Thus septic equations can be classified into the following three general types:

(1)   Special septic equations soluble by radicals—namely equations with the cyclic, dihedral, and metacyclic groups of orders 7 (C_7), 14 (D_7), 21, and 42 (M_7). The groups of orders 7, 14, and 21, are all subgroups of the metacyclic group.

(2)   Septic equations with the simple Galois group $L(3,2)$. Such equations require elliptic but not hyperelliptic functions for their solution and are discussed in this section. The $L(3,2)$ group can also be a permutation group of degree 8 and is associated with the modular equation of degree 8 (see Section 7.1). The theory for solving septic equations with the $L(3,2)$ Galois group is closely related to the theory for solving the general quintic equation (Chapters 6 and 7) although necessarily significantly more complicated.

(3)   General septic equations with the alternating or symmetric Galois groups A_7 or S_7. Such equations, like the general sextic equation (Section 8.1), require hyperelliptic functions and associated theta functions of genus 3 (Section 4.4) for their solution. General septic equations do not appear to have been studied specifically by the nineteenth century mathematicians studying the

solution of algebraic equations, apparently because solution of the general sextic equation was already at the limits of feasibility in that noncomputer era.

The alternating group A_5 of order 60 used for solution of the general quintic equation can be modelled geometrically by the proper rotations of a regular icosahedron, i.e., the symmetry point group I (Sections 2.2 and 2.3). The ability of $A_5 \equiv I$ to function as an even permutation group on five objects is then represented by the permutations of the five suboctahedra generated from the midpoints of the 30 edges of the underlying icosahedron (Figure 2–5).[9] Figure 8–1 depicts analogous geometrical models for the $L(3,2)$ group of order 168 which reflect its function as a permutation group of degree 7 for the special septic equation and as a permutation group of degree 8 for the modular equation of degree 8 as follows:

Degree 7: An equilateral triangle with its three altitudes and an inscribed circle forms a 7-point 7-line geometry presented in D_3 symmetry. The permutations of the 7 vertex labels which preserve the 7 collineations (152, 263, 173, 146, 247, 345, 567) form the $L(3,2)$ group. Note that in this presentation the inscribed circle is treated on an equal basis with the six straight lines forming the three edges and the three altitudes of the triangle.

Degree 8: The permutations of the vertex labels of a cuboid (rectangular box) of D_2 point group symmetry which give non-superimposable cuboids form the $L(3,2)$ group.

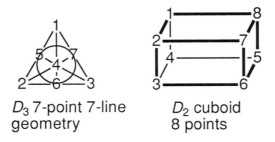

D_3 7-point 7-line geometry

D_2 cuboid 8 points

Figure 8–1: Geometrical models of the $L(3,2)$ group as a permutation group on 7 points and on 8 points.

---

[9]F. J. Budden, *The Fascination of Groups*, Cambridge, 1972, pp. 272–277, p. 411, and pp. 495–496.

Beyond the Quintic Equation

The $L(3,2)$ group has an associated series of polynomials connected by their transvectants analogous to the polyhedral polynomials discussed in Sections 2.5 and 2.6 (e.g., equations 2.5–9 defining f, T, and H for the icosahedron). However, in the case of the $L(3,2)$ group these polynomials are ternary forms with three homogeneous variables rather than the binary forms found in the polyhedral polynomials discussed in Section 2.5.

The polynomials associated with the $L(3,2)$ group were developed[10,11,12] in connection with the seventh order transformation of elliptic functions relating to the modular equation of degree 8. The homogeneous ternary quartic

$$f = \lambda^3 \mu + \mu^3 \nu + \nu^3 \lambda \qquad (8.2-1).$$

defines a fourth order curve by the relation $f = 0$. The Hessian of equation 8.2–1 leads to a ternary sextic ∇, i.e.,

$$\nabla = \frac{1}{54} \begin{vmatrix} \frac{\partial^2 f}{\partial \lambda^2} & \frac{\partial^2 f}{\partial \lambda \partial \mu} & \frac{\partial^2 f}{\partial \lambda \partial \nu} \\ \frac{\partial^2 f}{\partial \mu \partial \lambda} & \frac{\partial^2 f}{\partial \mu^2} & \frac{\partial^2 f}{\partial \mu \partial \nu} \\ \frac{\partial^2 f}{\partial \nu \partial \lambda} & \frac{\partial^2 f}{\partial \nu \partial \mu} & \frac{\partial^2 f}{\partial \nu^2} \end{vmatrix} = 5\lambda^2\mu^2\nu^2 - (\lambda^5\nu + \nu^5\mu + \mu^5\lambda) \qquad (8.2-2)$$

Related methods can be used to define C and K of degrees 14 and 21 as follows:

---

[10] F. Klein, Über die Transformation siebenter Ordnung der elliptischen Funktionen, *Math. Ann.*, **14**, 428–471 (1879); *Gesammelte Mathematische Abhandlungen*, Springer Verlag, Berlin, 1923, Vol. 3, pp. 90–136.
[11] P. Gordan, Über die typische Darstellung der ternären biquadratischen Form $f = x_1^3 x_2 + x_2^3 x_3 + x_3^3 x_1$, *Math. Ann.*, **14**, 359–378 (1880).
[12] H. Weber, *Lehrbuch der Algebra*, Vieweg, Braunschweig, 1898, Volume 2, §§ 131–140.

$$C = \frac{1}{9} \begin{vmatrix} \frac{\partial^2 f}{\partial \lambda^2} & \frac{\partial^2 f}{\partial \lambda \partial \mu} & \frac{\partial^2 f}{\partial \lambda \partial \nu} & \frac{\partial \nabla}{\partial \lambda} \\ \frac{\partial^2 f}{\partial \mu \partial \lambda} & \frac{\partial^2 f}{\partial \mu^2} & \frac{\partial^2 f}{\partial \mu \partial \nu} & \frac{\partial \nabla}{\partial \mu} \\ \frac{\partial^2 f}{\partial \nu \partial \lambda} & \frac{\partial^2 f}{\partial \nu \partial \mu} & \frac{\partial^2 f}{\partial \nu^2} & \frac{\partial \nabla}{\partial \nu} \\ \frac{\partial \nabla}{\partial \lambda} & \frac{\partial \nabla}{\partial \mu} & \frac{\partial \nabla}{\partial \nu} & 0 \end{vmatrix}$$

$$= \Sigma \lambda^{14} - 34\, \lambda\mu\nu \Sigma \lambda^{10}\nu - 250\, \lambda\mu\nu \Sigma \lambda^8 \mu^3 + 375\, \lambda^2 \mu^2 \nu^2 \Sigma \lambda^6 \mu^2$$
$$-126\, \lambda^3 \mu^3 \nu^3 \Sigma \lambda^3 \mu^2 + 18\, \Sigma \lambda^7 \mu^7 \qquad (8.2\text{--}3)$$

$$K = \nabla = \frac{1}{14} \begin{vmatrix} \frac{\partial f}{\partial \lambda} & \frac{\partial \nabla}{\partial \lambda} & \frac{\partial C}{\partial \lambda} \\ \frac{\partial f}{\partial \mu} & \frac{\partial \nabla}{\partial \mu} & \frac{\partial C}{\partial \mu} \\ \frac{\partial f}{\partial \nu} & \frac{\partial \nabla}{\partial \nu} & \frac{\partial C}{\partial \nu} \end{vmatrix}$$

$$= \Sigma \lambda^{21} - 7\, \lambda\mu\nu \Sigma \lambda^{17}\mu + 217\, \lambda\mu\nu \Sigma \lambda^{15}\mu^3 + 1638\, \lambda\mu\nu \Sigma^{10}\mu^8$$
$$- 308\, \lambda^2 \mu^2 \nu^2 \Sigma \lambda^{13}\mu^2 - 6279\, \lambda^2 \mu^2 \nu^2 \Sigma \lambda^9 \mu^6 + 637\, \lambda^3 \mu^3 \nu^3 \Sigma \lambda^9 \mu^3$$
$$+ 4018\, \lambda^3 \mu^3 \nu^3 \Sigma \lambda^{10}\mu^2 - 10010\, \lambda^4 \mu^4 \nu^4 \Sigma \lambda^5 \mu^4 + 7007\, \lambda^5 \mu^5 \nu^5 \Sigma \lambda^5 \nu$$
$$- 57(\lambda^{14}\mu^7 + \mu^{14}\nu^7 + \nu^{14}\lambda^7) - 289(\lambda^{14}\nu^7 + \mu^{14}\lambda^7 + \nu^{14}\mu^7) + 3432\, \lambda^7 \mu^7 \nu^7$$
$$(8.2\text{--}4)$$

The summation sign, Σ, in equations 8.2–3 and 8.2–4 means a sum of the three terms obtained from the first term by cyclic permutations of λ, μ, and ν; these equations are taken directly from Gordan's paper.[11] The homogeneous ternary polynomials ∇, C, and K satisfy a degree 42 identity

$$1728\, (-\nabla)^7 - C^3 + K^2 \equiv 0 \qquad (8.2\text{--}5)$$

closely related to the degree 60 icosahedral identity (compare equation 2.5–12c)

$$1728\, f^5 - H^3 - T^2 \equiv 0 \qquad (8.2\text{--}6).$$

Radford[13] has discussed the solution of septic equations with the $L(3,2)$ Galois group. Such equations are expressed by both Klein[10] and Radford[13] in the general form

$$c^7 + 7/2(-1 \mp \sqrt{-7})\nabla c^4 - 7\left(\frac{5 \mp \sqrt{-7}}{2}\right)\nabla^2 c - C = 0 \qquad (8.2\text{--}7)$$

in which the parameters ∇ and C are defined by equations 8.2–2 and 8.2–3, respectively. The complex coefficients in equation 8.2–7 can be eliminated by a suitable change in variable although this does not appear to have been worked out in detail, at least in the 19th century mathematical literature. Equation 8.2–7 resembles one form of the Brioschi quintic equation, namely

$$t^5 - 10ft^3 + 45f^2 t - T = 0 \qquad (8.2\text{--}8)$$

in which ∇ and C in equation 8.2–7 play analogous roles to f and T, respectively, in equation 8.2–8. The octic (degree 8) equation corresponding to equation 8.2–7, which also has an $L(3,2)$ Galois group, but now acting on eight roots corresponding to the cuboid vertices in Figure 8–1, has the form

$$\delta^8 - 14\nabla\delta^6 + 63\nabla^2\delta^4 - 70\nabla^3\delta^2 - K\delta - 7\nabla^4 = 0 \qquad (8.2\text{--}9)$$

This is an octic analogue of the following Jacobi sextic (compare equation 6.4–1):

$$s^6 - 10fs^3 + Hs + 5f^2 = 0 \qquad (8.2\text{--}10)$$

Radford[13] has expressed the roots of the octic equation (8.2–9) in product or sums related to theta series (Section 4.3) analogous to the formulas given in Section 6.6 for the Jacobi sextic (8.2–10) derived from the general quintic equation.

8.3 The General Algebraic Equation of Any Degree

Jordan[14] has shown that any algebraic equation can be solved using modular functions. The basis of such solutions involves Thomae's

---

[13] E. M. Radford, On the Solution of Certain Equations of the Seventh Degree, *Quart. J. Math.*, **30**, 263–306 (1898).

[14] C. Jordan, *Traité des Substitutions et des Équations Algébriques*, Gauthiers-Villars, Paris, 1870.

formulas[15,16,17]. The following theorem discussed by Umemura[18] provides a basis for expressing the roots of any algebraic equation by higher genus theta functions (Section 4.4):

Theorem 8.3–1:

$$f(x) = a_0 x^n + a_1 x^{n-1} + \ldots a_n = 0 \text{ with } a_0 \neq 0 \quad (8.3\text{–}1)$$

be an algebraic equation irreducible over a certain subfield of the complex numbers. Then a root of this equation, x_k, can be expressed by the following equation involving theta functions of *zero argument*:

$$x_k = \left[\theta\begin{pmatrix}1\,0\,\ldots 0\\0\ldots\,0\,0\end{pmatrix}(\Omega)\right]^4 \left[\theta\begin{pmatrix}1\,1\,0\ldots 0\\0\ldots 0\,0\,0\end{pmatrix}(\Omega)\right]^4 + \left[\theta\begin{pmatrix}0\ldots 0\\0\ldots 0\end{pmatrix}(\Omega)\right]^4 \left[\theta\begin{pmatrix}010\ldots 0\\000\ldots 0\end{pmatrix}(\Omega)\right]^4$$

$$- \frac{\left[\theta\begin{pmatrix}00\ldots 0\\10\ldots 0\end{pmatrix}(\Omega)\right]^4 \left[\theta\begin{pmatrix}01\,0\,\ldots 0\\10\ldots\,0\,0\end{pmatrix}(\Omega)\right]^4}{2\left[\theta\begin{pmatrix}1\,0\,\ldots 0\\0\ldots\,0\,0\end{pmatrix}(\Omega)\right]^4 \left[\theta\begin{pmatrix}1\,1\,\,0\,\ldots 0\\0\ldots\ldots\ldots 0\end{pmatrix}(\Omega)\right]^4} \quad (8.3\text{–}2)$$

In equation 8.3–2 Ω is the period matrix derived from one of the following hyperelliptic integrals:

$$u(a) = \int_1^a \frac{dx}{\sqrt{x(x-1)f(x)}} \qquad \text{for } n \text{ odd (i.e., odd degree of } f(x)) \quad (8.3\text{–}3a)$$

or

---

[15] J. Thomae, Beitrag zur Bestimmung von $\theta(0,0,\ldots,0)$ durch die Klassenmoduln algebraischer Functionen, *J. für Math*, **71**, 201–222 (1869).
[16] F. Lindemann, Über die Auflösung algebraischer Gleichungen durch transcendente Functionen I, II, *Göttingen Nach.*, 245–248 (1894); 292–298 (1892),
[17] D. Mumford, *Tata Lectures on Theta I*, Birkhäuser, Boston, 1983.
[18] H. Umemura, Resolution of Algebraic Equations by Theta Constants, in *Tata Lectures on Theta II*, D. Mumford, ed., pp. 3.261–3.272, Birkhäuser, Boston, 1984.

$$u(a) = \int_1^a \frac{dx}{\sqrt{x(x-1)(x-2)f(x)}} \quad \text{for } n \text{ even (i.e., even degree of } f(x)) \quad (8.3\text{--}3b)$$

This theorem applies to algebraic equations in any form without the need to use Tschirnhausen or other transformations to bring the original algebraic equation into a special form such as the Brioschi or Bring-Jerrard normal forms of the quintic equation. However, application of this theorem in practice is very difficult because of the complexity of the relevant hyperelliptic integrals and higher genus theta functions.